MORE GREAT FORMULAS EXPLAINED

-

PHYSICS, MATHEMATICS

METIN BEKTAS

ISBN: 9781520975702

DEDICATION

This book is dedicated to my family.

CONTENTS

Part I: Physics

- **Law Of The Lever**

Often times when doing physics we simply say "a force is acting on a body" without specifying which point of the body it is acting on. This is basically point-mass physics. We ignore the fact that the object has a complex three-dimensional shape and assume it to be a single point having a certain mass. Sometimes this is sufficient, other times we need to go beyond that. And this is where the concept of torque comes in.

Let's define what is meant by torque. Assume a force F (in N) is acting on a body at a distance r (in m) from the axis of rotation. This distance is called the lever arm. Take a look at the image below for an example of such a set up.

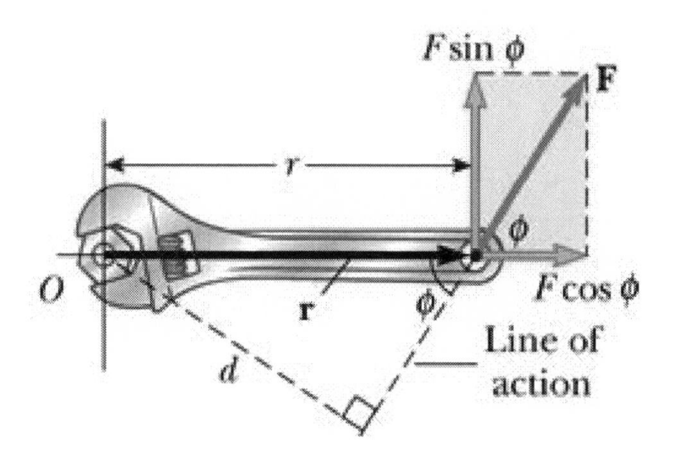

Relevant for the rotation of the body is only the force component perpendicular to the lever arm, which we will denote by F'. If given the angle Φ between the force and the lever arm (as shown in the image), we can easily compute the relevant force component by:

$$F' = F \cdot \sin(\Phi)$$

For example, if the total force is F = 50 N and it acts at an angle of $\Phi = 45°$ to the lever arm, only the the component $F' = 50\ N \cdot \sin(45°) \approx 35\ N$ will work to rotate the body. So you can see that sometimes it makes sense to break a force down into its components. But this shouldn't be cause for any worries, with the above formula it can be done quickly and painlessly.

With this out of the way, we can define what torque is in one simple sentence: Torque T (in Nm) is the product of the lever arm r and the force F' acting perpendicular to it. In form of an equation the definition looks like this:

$$T = r \cdot F'$$

In quantitative terms we can interpret torque as a measure of rotational push. If there's a force acting at a large distance from the axis of rotation, the rotational push will be strong. However, if one and the same force is acting very close to said axis, we will see hardly any rotation. So when it comes to rotation, force is just one part of the picture. We also need to take into consideration where the force is applied.

Let's compute a few values before going to the extremely useful law of the lever.

We'll have a look at the wrench from the image. Suppose the wrench is $r = 0.2$ m long. What's the resulting torque when applying a force of $F = 80$ N at an angle of $\Phi = 70°$ relative to the lever arm?

To answer the question, we first need to find the component of the force perpendicular to the lever arm.

$F' = 80\ N \cdot \sin(70°) \approx 75.18\ N$

Now onto the torque:

$T = 0.2\ m \cdot 75.18\ N \approx 15.04\ Nm$

If this amount of torque is not sufficient to turn the nut, how could we increase that? Well, we could increase the force F and at the same time make sure that it is applied at a 90° angle to the wrench. Let's assume that as a measure of last resort, you apply the force by standing on the wrench. Then the force perpendicular to the lever arm is just your gravitational pull:

$F' = F = m \cdot g$

Assuming a mass of m = 75 kg, we get:

$F' = 75\ kg \cdot 9.81\ m/s^2 = 735.75\ N$

With this not very elegant, but certainly effective technique, we are able to increase the torque to:

$T = 0.2\ m \cdot 735.75\ N = 147.15\ Nm$

That should do the trick. If it doesn't, there's still one option left and that is using a longer wrench. With a longer wrench you can apply the force at a greater distance to the axis of rotation. And with r increased, the torque T is increased by the same factor.

As you can see, calculating torque is not a big deal. But what's the use? The law of the lever, that's what. Imagine a beam sitting on a fulcrum. We apply one force F'(1) = 20 N on the left side at a distance of r(1) = 0.1 m from the fulcrum and another force F'(2) = 5 N on the right side at a distance of r(2) = 0.2 m. In which direction, clockwise or anti-clockwise, will the beam move?

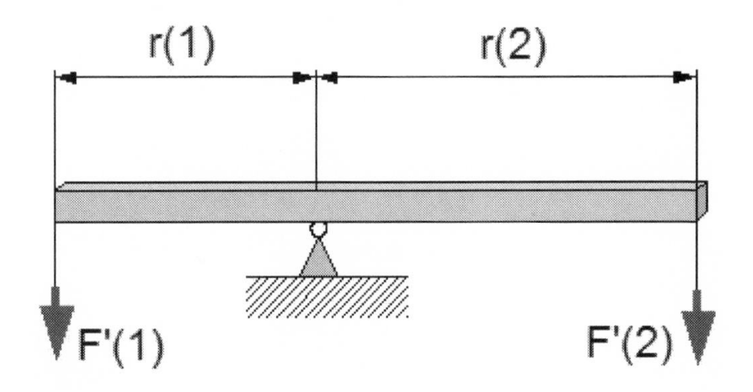

To find that out we can take a look at the corresponding torques. The torque on the left side is:

$$T(1) = 0.1 \text{ m} \cdot 20 \text{ N} = 2 \text{ Nm}$$

For the right side we get:

$$T(2) = 0.2 \cdot 5 \text{ N} = 1 \text{ Nm}$$

So the rotational push caused by force 1 (left side) exceeds that of force 2 (right side). Hence, the beam will turn anti-clockwise. If we don't want that to happen and instead want to achieve equilibrium, we need to increase force 2 to F'(2) = 10 N. In this case the torques would be equal and the opposite rotational pushes would cancel each other. So in general, this equation needs to be satisfied to achieve a state of equilibrium:

$r(1) \cdot F'(1) = r(2) \cdot F'(2)$

This is the law of the lever in its simplest form. Let's see how and where we can apply it.

A great example for the usefulness of the law of the lever is provided by cranes. On one side, let's set $r(1) = 30$ m, it lifts objects. Since we don't want it to fall over, we stabilize the crane using a 20,000 kg concrete block at a distance of $r(2) = 2$ m from the axis. What is the maximum mass we can lift with this crane?

First we need to compute the gravitational force of the concrete block.

$F'(2) = 20{,}000 \text{ kg} \cdot 9.81 \text{ m/s}^2 = 196{,}200 \text{ N}$

Now we can use the law of the lever to find out what maximum force we can apply on the opposite site:

$r(1) \cdot F'(1) = r(2) \cdot F'(2)$

$30 \text{ m} \cdot F'(1) = 2 \text{ m} \cdot 196{,}200 \text{ N}$

$30 \text{ m} \cdot F'(1) = 392{,}400 \text{ Nm}$

Divide by 30 m:

$F'(1) = 13{,}080 \text{ N}$

As long as we don't exceed this, the torque caused by the concrete block will exceed that of the lifted object and the crane will not fall over. The maximum mass we can lift is now easy to find. We use the formula for the gravitational force one more time:

$13{,}080\ N = m \cdot 9.81\ m/s^2$

Divide by 9.81:

$m \approx 1330\ kg$

To lift even heavier objects, we need to use either a heavier concrete block or put it at a larger distance from the axis.

The law of the lever shows why we can interpret a lever as a tool to amplify forces. Suppose you want use a force of F'(1) = 100 N to lift a heavy object with the gravitational pull F'(2) = 2000 N. Not possible you say? With a lever you can do this by applying the smaller force at a larger distance to the axis and the larger force at a shorter distance.

Suppose the heavy object sits at a distance r(2) = 0.1 m to the axis. At what what distance r(1) should we apply the 100 N to be able to lift it? We can use the law of the lever to find the minimum distance required.

$r(1) \cdot 100\ N = 0.1\ m \cdot 2000\ N$

$r(1) \cdot 100\ N = 200\ Nm$

$r(1) = 2\ m$

So as long as we apply the force at a distance of over 2 m, we can lift the object. We effectively amplified the force by a factor of 20. Scientists believe that the principle of force amplification using levers was already used by the Egyptians to build the pyramids. Given a long enough lever, we could lift basically anything even with a moderate force.

In the next section we will see another interesting application of the torque concept, so we're not done with it just yet. It will lead to a very neat formula all car makers must know.

- **Sliding and Overturning**

In this section we will take a look at car performance in curves. Of central importance for our considerations is the centrifugal force. Whenever a body is moving in a curved path, this force comes into play. You probably felt it many times in your car. It is the force that tries to push you out of a curve as you go through it.

The centrifugal force C (in N) depends on three factors: the velocity v (in m/s) of the car, its mass m (in kg) and the radius r (in m) of the curve. Given these quantities, we can easily compute the centrifugal force using this formula:

$$C = m \cdot v^2 / r$$

Note the quadratic dependence on speed. If you double the car's speed, the centrifugal force quadruples. With this force acting, there must be a counter-force to cancel it for the car not to slide. This force is provided by the sideways friction of the tires. The frictional force F (in N) can be calculated from the so called coefficient of friction μ (dimensionless), the car mass m and the gravitational acceleration g (in m/s^2).

$$F = \mu \cdot m \cdot g$$

The coefficient of friction depends mainly on the road type and condition. On dry asphalt we can set $\mu \approx 0.8$, on wet asphalt $\mu \approx 0.6$, on snow $\mu \approx 0.2$ and on ice $\mu \approx 0.1$. At low speeds the frictional force exceeds the centrifugal force and the car will be able to go through the curve without any problems. However, as we increase the velocity, so does the centrifugal force and at a certain critical velocity the forces cancel each other out. Any increase in speed from this point on will result in the car sliding.

We can compute the critical speed s (in m/s) by equating the expressions for the forces:

$$m \cdot s^2 / r = \mu \cdot m \cdot g$$

$$\mathbf{s = sqrt(\mu \cdot r \cdot g)}$$

This is the speed at which the car begins to slide. Note that there's no dependence on mass anymore. Since both the centrifugal as well as the frictional force grow proportionally to the car's mass, it doesn't play a role in determining the critical speed for sliding. All that's left in terms of variables is the coefficient of friction (lower friction, lower critical speed) and the radius of the curve (smaller radius, more narrow curve, smaller critical speed).

However, sliding is not the only problem that can occur in curves. Under certain circumstances a car can also overturn. Again the centrifugal force is the culprit. Assuming the center of gravity (in short: CG) of the car is at a height of h (in m), the centrifugal force will produce a torque T acting to overturn the car:

$$T = h \cdot C = m \cdot v^2 \cdot h / r$$

On the other hand, there's the weight of the car giving rise to an opposing torque T' that grows with the width w (in m) and mass m of the car:

$$T' = 0.5 \cdot m \cdot g \cdot w$$

At low speeds, the torque caused by the centrifugal force will be lower than the one caused by the gravitational pull. But at a certain critical speed o (in m/s), the torques will cancel and any further increase in speed will result in the car overturning. Equating the expressions, we get:

$m \cdot o^2 \cdot h / r = 0.5 \cdot m \cdot g \cdot w$

$\mathbf{o = sqrt\ (0.5 \cdot r \cdot g \cdot w / h)}$

Aside from the coefficient of friction, the determining factor here is the ratio of width to height. The larger it is, the harder it will be for the centrifugal force to overturn the car. This is why lowering a car when intending to go fast makes sense. If you lower the CG while keeping the width the same, the ratio w / h, and thus the critical speed for overturning, will increase.

Let's look at some examples before drawing a final conclusion from these truly great formulas.

According to caranddriver.com the center of gravity of a 2014 BMW 435i is $h = 0.5$ m above the ground. The width of the car is about $w = 1.8$ m. Calculate the critical speed for sliding and overturning in a curve of radius $r = 300$ m on a dry asphalt road ($\mu \approx 0.8$).

Nothing to do but to apply the formulas:

$s = sqrt\ (0.8 \cdot 300\ m \cdot 9.81\ m/s^2)$

$s \approx 49\ m/s \approx 175\ km/h \approx 108\ mph$

So with normal driving behavior you certainly won't get anywhere near sliding. But note that sudden steering in a curve can cause the radius of the your car's path to be considerably lower than the actual curve radius.

Onto the critical overturning speed:

o = sqrt (0.5 · 300 m · 9.81 m/s² · 3.6)

o ≈ 73 m/s ≈ 262 km/h ≈ 162 mph

Not even the great Michael Schumacher could bring this car to overturn.

How would the critical speeds change if we drove the 2014 BMW 435i through the same curve on an icy road? In this case the coefficient is considerably lower (μ ≈ 0.1). For the critical sliding speed we get:

s = sqrt (0.1 · 300 m · 9.81 m/s²)

s ≈ 17 m/s ≈ 62 km/h ≈ 38 mph

So even this sweet sport car is in danger of sliding relatively quickly under these conditions. What about the overturning speed? Well, it has nothing to do with the friction of the tires, so it will still be at 73 m/s.

Sliding does not always end in an accident. While sliding, a car will usually increase the radius of its path and at the same time decrease its speed. Both of these effects lower the centrifugal force and thus can bring the car back into its normal driving mode before an accident occurs. On the other hand, once a car starts overturning, there's no going back.

This is why when designing a car, you want it to slide before it overturns, so the critical sliding speed should be lower than the critical speed for overturning. Using the formulas, we can find a critical CG height H (in m) by equating the critical speeds:

$$s = o$$

$$H = 0.5 \cdot w / \mu$$

As long as the CG is below this height, the car will start sliding before overturning. Since we saw in the example that the difference between the critical speeds increases as the road conditions worsen, it is sufficient to make sure that the above equation holds true for dry asphalt. Setting $\mu \approx 0.8$, we get:

$$\mathbf{H \approx 0.63 \cdot w}$$

So as a rule of thumb, the height of the CG should be kept below 60 % of the width. Of course accurately determining the CG not something you can do with a ruler. But it can be estimated by measuring the weights on each tire in a horizontal position and the weights on the front tires after raising the front. With this done, you can input these measurements into the online CG calculator featured on robrobinette.com.

- **Maximum Car Speed**

How do you determine the maximum possible speed your car can go? Well, one rather straight-forward option is to just get into your car, go on the Autobahn and push down the pedal until the needle stops moving. The problem with this option is that there's not always an Autobahn nearby. So we need to find another way.

Luckily, physics can help us out here. You probably know that whenever a body is moving at constant speed, there must be a balance of forces in play. The force that is aiming to accelerate the object is exactly balanced by the force that wants to decelerate it. Our first job is to find out what forces we are dealing with.

Obvious candidates for the retarding forces are ground friction and air resistance. However, in our case looking at the latter is sufficient since at high speeds, air resistance becomes the dominating factor. This makes things considerably easier for us. So how can we calculate air resistance?

To compute air resistance we need to know several inputs. One of these is the air density D (in kg/m^3), which at sea level has the value $D = 1.25\ kg/m^3$. We also need to know the projected area A (in m^2) of the car, which is just the product of width times height. Of course there's also the dependence on the velocity v (in m/s) relative to the air. The formula for the drag force is:

$$F = 0.5 \cdot c \cdot D \cdot A \cdot v^2$$

with c (dimensionless) being the drag coefficient. This is the one quantity in this formula that is tough to determine. You probably don't know this value for your car and there's a good chance you will never find it out even if you try. In general, you want to have this value as low as possible.

On ecomodder.com you can find a table of drag coefficients for many common modern car models. Excluding prototype models, the drag coefficient in this list ranges between $c = 0.25$ for the Honda Insight to $c = 0.58$ for the Jeep Wrangler TJ Soft Top. The average value is $c = 0.33$. In first approximation you can estimate your car's drag coefficient by placing it in this range depending on how streamlined it looks compared to the average car.

With the equation: power equals force times speed, we can use the above formula to find out how much power (in W) we need to provide to counter the air resistance at a certain speed:

$$P = F \cdot v = 0.5 \cdot c \cdot D \cdot A \cdot v^3$$

Of course we can also reverse this equation. Given that our car is able to provide a certain amount of power P, this is the maximum speed v we can achieve:

$$\mathbf{v = (2 \cdot P / (c \cdot D \cdot A))^{1/3}}$$

From the formula we can see that the top speed grows with the third root of the car's power, meaning that when we increase the power eightfold, the maximum speed doubles. So even a slight increase in top speed has to be bought with a significant increase in energy output.

Note the we have to input the power in the standard physical unit watt rather than the often used unit horsepower. Luckily the conversion is very easy, just multiply horsepower with 746 to get to watt.

Let's put the formula to the test.

I drive a ten year old Mercedes C180 Compressor. According the Mercedes-Benz homepage, its drag coefficient is $c = 0.29$ and its power $P = 143\ HP \approx 106{,}680\ W$. Its width and height is $w = 1.77\ m$ and $h = 1.45\ m$ respectively. What is the maximum possible speed?

First we need the projected area of the car:

$A = 1.77\ m \cdot 1.45\ m \approx 2.57\ m^2$

Now we can use the formula:

$v = (\ 2 \cdot 106{,}680\ /\ (0.29 \cdot 1.25 \cdot 2.57)\ )^{1/3}$

$v \approx 61.2\ m/s \approx 220.3\ km/h \approx 136.6\ mph$

From my experience on the Autobahn, this seems to be very realistic. You can reach 200 Km/h quite well, but the acceleration is already noticeably lower at this point.

If you ever get the chance to visit Germany, make sure to rent a ridiculously fast sports car (you can rent a Porsche 911 Carrera for as little as 200 $ per day) and find a nice section on the Autobahn with unlimited speed. But remember: unless you're overtaking, always use the right lane. The left lanes are reserved for overtaking. Never overtake on the right side, nobody will expect you there. And make sure to check the rear-view mirror often. You might think you're going fast, but there's always someone going even faster. Let them pass. Last but not least, stay focused and keep your eyes on the road. Traffic jams can appear out *of nowhere and you don't want to end up in the back of a truck at these speeds.*

The fastest production car at the present time is the Bugatti Veyron Super Sport. Is has a drag coefficient of c = 0.35, width w = 2 m, height h = 1.19 m and power P = 1200 HP = 895,200 W. Let's calculate its maximum possible speed:

$v = (2 \cdot 895{,}200 / (0.35 \cdot 1.25 \cdot 2 \cdot 1.19))^{1/3}$

$v \approx 119.8\ m/s \approx 431.3\ km/h \approx 267.4\ mph$

Does this seem unreasonably high? It does. But the car has actually been recorded going 431 Km/h, so we are right on target. If you'd like to purchase this car, make sure you have 4,000,000 $ in your bank account.

To conclude this section, let's do a very rough estimate on how long it takes a car to reach its maximum speed v. At this speed the car's kinetic energy is $E = 0.5 \cdot m \cdot v^2$. If the car continuously produces the maximum power P, this is how long it takes to provide the above amount of energy:

$$\mathbf{t = 0.5 \cdot m \cdot v^2 / P}$$

with t being in seconds. Note that since the speed grows with the third root of the power, the time it takes to reach maximum speed should be inversely proportional to the third root of the power. So if you increase the power eightfold, it will take only half as long to go to top speed. Also note that here the mass of the car is also a factor.

We determined that for the Mercedes C180 Compressor the top speed is $v = 61.2$ m/s. Given the power $P = 143$ HP $\approx$ 106,680 W and the mass $m = 1,600$ kg, estimate how long it takes to reach the maximum speed.

We apply the formula:

$t = 0.5 \cdot 1{,}600 \cdot 61.2^2 / 106{,}680$

$t \approx 28.1$ s

This is of course assuming that we could constantly produce the maximum power (which is not the case). But the value nevertheless seems quite realistic considering the car can go to 100 Km/h (about half the top speed) in 9 seconds and the acceleration would be significantly slower in the second half.

What about the Bugatti Veyron Super Sport? How long does it take to get it up to its theoretical top speed of $v = 119.8$ m/s? The mass of the car is $m = 1900$ kg, hence we get:

$t = 0.5 \cdot 1{,}800 \cdot 119.8^2 / 895{,}200$

$t \approx 15.2$ s

So not only do we reach twice the top speed of the Mercedes C180 Compressor, we get there more than 10 seconds earlier. By the way, the average acceleration during this would be:

$a = v / t = 119.8 / 15.2 \approx 7.9\ m/s^2$

which is comparable to what we'd experience in free fall. And this is just the average. It actually goes from 0 to 100 Km/h (27.8 m/s) in only 2.2 seconds, which translates into an acceleration of:

$a = 27.8 / 2.2 \approx 12.6\ m/s^2$

Try these calculations for your own car. If you're lucky you'll find its drag coefficient online. If not, just estimate it from range provided in this section. You shouldn't be off by too much.

- **Range Continued**

In the first volume of this book we had a look at how far an object thrown from the ground at a certain velocity and angle gets. We'll take yet another look at range, but this time for a different set up. Assume we throw an object horizontally at a speed of v (in m/s) from the height h (in m) above the ground. What horizontal distance will it cover before hitting the ground?

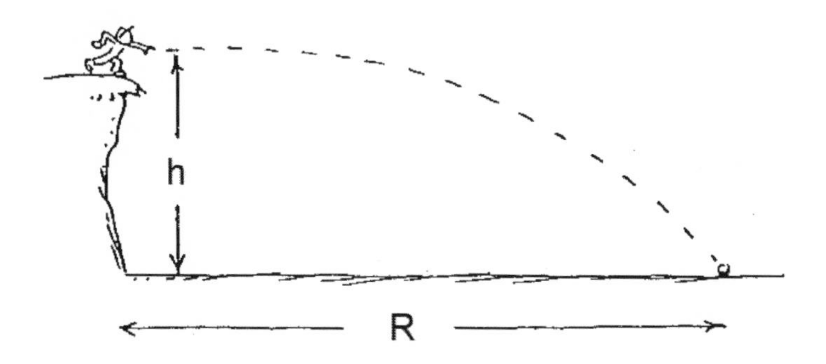

Well, in x-direction (which is the horizontal axis), the object will keep on moving at the velocity v when neglecting air resistance:

$x = v \cdot t$

In y-direction it is subject to the constant gravitational acceleration $g = 9.81\ m/s^2$. So the height will decrease according to this formula:

$y = h - 0.5 \cdot g \cdot t^2$

We can get rid of time by using $t = x / v$ and plugging that into the second formula.

$y = h - 0.5 \cdot g \cdot x^2 / v^2$

This is the coordinate form of the parabolic path the object takes. We want to know at what value $x = R$ it impacts the ground ($y = 0$). With the above equation this is easy to do. We set:

$$0 = h - 0.5 \cdot g \cdot R^2 / v^2$$

And solve for the range:

$$\mathbf{R = v \cdot sqrt\ (2 \cdot h / g)}$$

In just a few steps we derived a great formula that allows us to compute the range from the initial velocity of the object and the initial height. Let's talk about the nature of the dependencies. The range grows linearly with velocity. If the velocity doubles, so does the range. As for the height, quadrupling it will also lead to a doubling in range.

Before we go on to the examples, let's take a closer look at the impact itself. We might be interested in knowing at what speed and at what angle the object impacts the ground. How can we compute these quantities? Again the key is to analyze the x- and y- direction separately.

As stated, the speed in x-direction will remain equal to the initial speed, so $v(x) = v$. And according to the formula: speed equals acceleration multiplied by time, the speed in y-direction after time t will be: $v(y) = g \cdot t$.

Let's get rid of t. How long does it take the object to hit the ground? This is easy. We already know that it takes the object the time $t = x / v$ to cover the horizontal distance x. So the time to impact is just:

$$t = R / v = sqrt\ (2 \cdot h / g)$$

Now we are able to compute the speed in y-direction at the time of impact. Inserting this expression for t into $v(y) = g \cdot t$ leads to:

$$v(y) = \text{sqrt}\ (2 \cdot g \cdot h)$$

Since the x- and y-velocities are at a right angle to each other (see image below), the total speed can be computed using Pythagoras' theorem:

$$w = \text{sqrt}\ (v(x)^2 + v(y)^2)$$

All that's left is inserting the appropriate expressions and we are left with a handy formula for calculating the total impact speed w (in m/s).

$$\mathbf{w = sqrt\ (v^2 + 2 \cdot g \cdot h)}$$

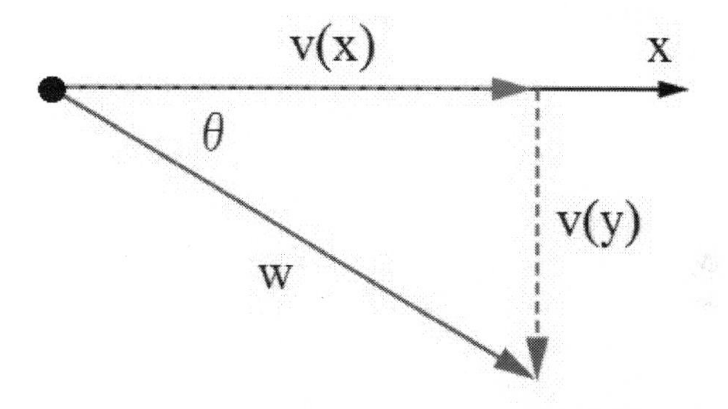

What about the angle of impact θ? For that we will need some trigonometry. Remember that in the tangent formula we use the ratio: opposite to adjacent. In our case the opposite is the speed in y-direction and the adjacent the speed in x-direction. Hence we can write:

$$\mathbf{\tan \theta = sqrt\ (2 \cdot g \cdot h) / v}$$

Note that the velocity v is not part of the square root. The relationship is somewhat complex, but in general we can conclude that the larger the ratio of initial height to initial speed is, the steeper the impact will be.

After all this work, let's go to the examples.

A ball rolls over the edge of a table of height $h = 1$ m at a speed of $v = 3$ m/s. At what horizontal distance to the edge of the table will it land? At what speed and angle does it impact the ground? First let's take a look the the range:

$R = 3 \text{ m/s} \cdot \text{sqrt}(2 \cdot 1 \text{ m} / 9.81 \text{ m/s}^2)$

$R \approx 1.35 \text{ m}$

So it will land at a distance of 1.35 m from the table. Now let's turn to the impact speed:

$w = \text{sqrt}((3 \text{ m/s})^2 + 2 \cdot 9.81 \text{ m/s}^2 \cdot 1 \text{ m})$

$w \approx 5.35 \text{ m/s}$

So the ball's speed increases significantly (by about 80 %) during the fall. The angle of impact is:

$\tan \theta = \text{sqrt}(2 \cdot 9.81 \text{ m/s}^2 \cdot 1 \text{ m}) / 3 \text{ m/s}$

$\tan \theta \approx 1.48$

Using the inverse function we get:

$\theta \approx \arctan(1.48) \approx 55.9°$

We hold a garden hose horizontally at a height of $h = 0.5$ m and turn on the water. The water jet hits the ground at a distance of $R = 1.5$ m. At what speed is the water exiting the hose? From the range formula we can set up an equation for the initial speed v using the given values (we'll ignore the units for sake of simplicity):

$1.5 = v \cdot sqrt(2 \cdot 0.5 / 9.81)$

$4 \approx v \cdot 0.32$

Divide by 0.32:

$v \approx 12.5$ m/s

Just after jumping out of the vehicle, the owner's brand new car goes over a cliff of height $h = 15$ m. The police determine from the impact crater that the angle of impact was about $\theta = 25°$. The owner insist that he was below the 100 Km/h limit at the time of the accident. Is he telling the truth?

Let's calculate the initial speed v and compare the result to the limit. From the formula for the angle of impact we can set up this equation:

$tan(25°) = sqrt\ (2 \cdot 9.81 \cdot 15) / v$

$0.47 \approx 17.16 / v$

Multiply by v:

$0.47 \cdot v \approx 17.16$

Now divide by 0.47:

$v \approx 36.5$ *m/s*

So this is the speed at which the car went over the cliff. How does that compare to the speed limit? Well, 36.5 m/s are 131.4 km/h, so he was significantly over the limit.

To conclude this section I'll give you another relevant formula you will hardly find in any other physics book. It is possible to compute the total distance covered by the object (= the length of the parabolic arc). Since the derivation requires a trip deep into the realms of integral calculus, I'll skip it entirely and go right to the formula for the arc length. Brace yourself:

$$\mathbf{s = sqrt\,(R^2 + 4 \cdot h^2) + (0.5 \cdot R^2 / h) \cdot arcsinh(2 \cdot h / R)}$$

with arcsinh being the inverse function of the hyperbolic sine function (your calculator should know this one). Granted, this is a monster. But it works. And once we know the total distance covered, we can go on to calculate the average speed <v>:

$$\langle v \rangle = s / t = s \cdot v / R$$

Let's give it a try.

Let's go back to the second example where we had water flowing out of a hose at a height of h = 0.5 m. The range was R = 1.5 m. What is the length of the water jet from hose to ground? We apply the monster formula:

$s = sqrt\,(3.25) + 2.25 \cdot arcsinh(0.67)$

$s \approx 3.21\ m$

Once you figure out how where the inverse hyperbolic sine function is on your calculator, this isn't actually so bad.

As you can see, even in such a simple and basic situation as free fall there's a lot of interesting physics and even some hard-core math involved. And it gets even worse once you want to include air resistance (don't try this at home!).

- **Escape Velocity**

If you throw an object upwards, gravity will force it to come back down quickly. After all, what goes up must come down. Unless it's going faster than the escape velocity, in which case it indeed will be able to escape a planet's or moon's gravitational prison. In this section we will derive and apply a formula that allows us to calculate said escape velocity (under the assumption of there being no atmosphere present).

To derive a formula we need to think about how much energy is needed to move an object from one point to another in a gravitational field. Most people here would say (if they'd say anything): just use this formula for the potential energy.

$E(pot) = m \cdot g \cdot h$

From this we can compute how much energy is needed to elevate an object of mass m (in kg) by a height of h (in m). Problem solved? Yes and no. This formula works as long as the change in height is not too great. For people, elevators, planes and even high-altitude balloons it works just fine. But for satellites and spaceships (as well as for calculating the escape velocity) it unfortunately fails. So we need something more precise.

Luckily, with Newton's formula for the gravitational force and some calculus, we can derive this formula that allows us to find out the energy needed to bring an object from the surface of a celestial body to a certain height:

$$\mathbf{E(pot) = G \cdot M \cdot m \cdot (1/R - 1/(R+h))}$$

with M (in kg) being the mass of the celestial body, R (in m) its radius and G the gravitational constant. Its value is: $G \approx 6.67 \cdot 10^{-11}$ Nm^2kg^2 throughout the universe and for all bodies. It can't get more constant than that, I suppose.

To calculate how much energy is needed to completely free an object of a gravitational field, we let the height go to infinity, which we express symbolically as such: $h \rightarrow \infty$. The second term in the above formula then just disappears, leaving this much nicer expression:

$$E(pot, \infty) = G \cdot M \cdot m / R$$

This is the energy we need to provide an object with so that it can leave a planet for good (assuming no atmosphere present). How does this translate into a velocity? Well, we know that an object with velocity v (in m/s) has the kinetic energy $E(kin) = 0.5 \cdot m \cdot v^2$. So we can find the escape velocity by equating the kinetic energy with the potential energy:

$$E(kin) = E(pot, \infty)$$

$$0.5 \cdot m \cdot v^2 = G \cdot M \cdot m / R$$

Solving for v we get:

$$\mathbf{v = sqrt (2 \cdot G \cdot M / R)}$$

This is the escape velocity. Note that it does not depend on the mass of the object that is to be shot into space in any way. All the parameters in the formula refer to the celestial body only. And it all comes down to the ratio of mass to radius. The greater this is, the higher the escape velocity will be.

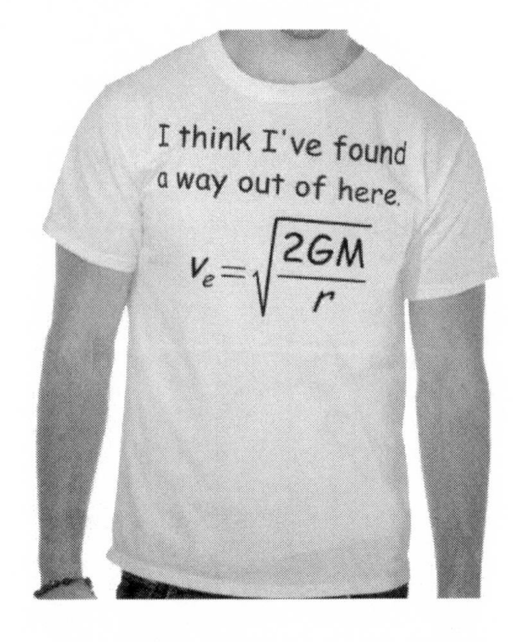

Let's derive an alternate form to the above equation that will allow us to go around having to deal with very low (G) or very high (M) numbers. From Newton's law of gravitation we can quickly show that a celestial body's gravitational acceleration at the surface (denoted by g and well-known for many bodies within our solar system) is:

$g = G \cdot M / R^2$

Thus we can write for the escape velocity:

v = sqrt (2 · g · R)

No unreasonably low or large numbers in there. It makes life easier when looking at bodies we already know. The same simplification we can do of course for the potential energy. With $G \cdot M = g \cdot R^2$ we get:

$E(pot) = m \cdot g \cdot R^2 \cdot (1/R - 1/(R+h))$

$E(pot, \infty) = m \cdot g \cdot R$

I would advise to use the latter formulas if the gravitational acceleration g is known (no need to reinvent the wheel, especially if the wheel has an unpractical ten to the power of minus eleven in it). Here are some values: $g = 9.81$ m/s² for earth, $g = 1.62$ for the moon, $g = 3.72$ for Mars and $g = 274$ for the sun.

Now onto the examples.

We want to bring a $m = 1200$ kg satellite into low earth orbit ($h = 1000$ km). Calculate the potential energy necessary for that. Additionally, compute how much energy is needed to get the same satellite out of earth's gravitational pull. The radius of earth is $R = 6400$ km.

Let's do the first part of the problem. Using the formula for E(pot) with $g = 9.81$ m/s² we get the following. Remember to input all lengths in meters.

$E(pot) \approx 10{,}181{,}190{,}000\ J \approx 10.18\ GJ$

This is roughly the energy you get from burning one ton of brown coal. What about the second part of the problem? Now we want to bring the same satellite out of earth's reach. Accordingly, we use the formula for E(pot, ∞):

$E(pot, \infty) = 75.340.800.000 \approx 75.34\ GJ$

So from an energetic point of view, we're only about one seventh there at a height of 1000 km.

Compute the escape velocity for earth ($g = 9.81\ m/s^2$ and $R = 6400$ Km) and compare it to that of the moon ($g = 1.62\ m/s^2$ and $R = 1740$ Km). Make sure to input all lengths in meters.

For earth we get:

$v = sqrt\ (2 \cdot 9.81\ m/s^2 \cdot 6{,}400{,}000\ m)$

$v \approx 11{,}206\ m/s \approx 11.2\ km/s \approx 6.9\ miles/s$

And for the moon:

$v = sqrt\ (2 \cdot 1.62\ m/s^2 \cdot 1{,}740{,}000\ m)$

$v \approx 2374\ m/s \approx 2.4\ km/s \approx 1.5\ miles/s$

So the escape velocity for the moon is only about one-fifth that of earth. Though 1.5 miles per second is still crazy fast.

Let's take a look at a collapsed star. We assume its mass to be $M = 3 \cdot M(sun) \approx 6 \cdot 10^{30}$ kg and its radius $R = 20$ m (which is realistic for a collapsed star). What is its escape velocity? And: do you notice something odd?

To solve this problem we need to go back to the highlighted formula for the escape velocity. We get:

$v = sqrt\ (2 \cdot 6.67 \cdot 10^{-11} \cdot 6 \cdot 10^{30} / 20)$

$v \approx 6.326.136.000\ m/s$

What's odd about this (other than that it's ridiculously high)? Well, the speed of light is about 300,000,000 m/s and the escape velocity in this case is greater than that. That's a problem! If you know anything about physics, it's that you cannot go faster than the speed of light. It is the speed limit for our universe. So nothing can escape the gravitational field of this body, not even light.

This is what physicist call a black hole and we know nothing about what's in there because no information can come out of it. For all we know, there might be two-headed dragons fighting giant space pirates in there. Not likely, I know, but prove me wrong.

For now, let's go back to earth. But at a later point in the book we'll be taking another look deep into space when talking about heat radiation and main sequence stars.

- **Cooling and Wind-Chill**

Most people associate Newton with the motion of bodies and gravitation. However, he also did considerable work in thermodynamics. In this section we will take a look at his law of cooling and the very informative conclusions we can draw from it.

What determines how fast an object cools? A main parameter here is the difference between the temperature of the object T (in °C) and the temperature U (in °C) of the surrounding. The bigger this difference is, the more heat energy the object loses to the environment per second. Another factor is the surface area S (in m^2) of the object. More exposed surface means greater heat loss.

Newton derived a simple formula that enables us to compute the heat loss per second P (in J/s = W), also called heat flow rate, from the quantities mentioned above.

$$\mathbf{P = \alpha \cdot S \cdot (T - U)}$$

Obviously there's one more factor involved. α is called the heat transfer coefficient and depends on the material surrounding the object (which is usually air or water). The coefficient is also influenced by the velocity of the surrounding material relative to the object. We will take a closer look at the implications of this later. For the most common situations you can find a suitable value for α in tables.

Note that despite calling it a law of cooling, the formula can also be applied when the temperature of the surrounding is lager than the temperature of the object. It's just as much a law of heating as it is a law of cooling. In this case P will be the heat gained per second.

Let's apply the formula real quick before moving on to another formula that will allow us to estimate the time it takes the object to reach the temperature of the surrounding.

We put a can of coke currently at room temperature $T = 25$ °C into a fridge with an inside temperature of $U = 5$ °C. The surface are of the can is approximately $S = 0.03\ m^2$. Assuming the heat transfer coefficient to be $\alpha = 10\ W/(m^2 K)$, which seems about right from what I can find in tables, how much heat energy does the coke can lose per second?

Let's plug in the values:

$P = 10 \cdot 0.03 \cdot 20 = 6\ J/s$

As the can drops in temperature, the rate of cooling will decrease. For example, once the can has reached a temperature of $T = 10$ °C, the heat flow rate has decreased to:

$P = 10 \cdot 0.03 \cdot 5 = 1.5\ J/s$

This demonstrates the exponential nature of the cooling process. The closer the object's temperature is to the surrounding, the slower it will cool, only gradually reaching the temperature of the surrounding.

Assume we want to speed up the cooling in the fridge and for this reason install a fan that gently circulates the air within the fridge. If the velocity of air is v = 1 m/s, the heat transfer coefficient increases to approximately $\alpha = 14\ W/(m^2\ K)$*. How does that change the initial heat flow rate?*

$P = 14 \cdot 0.03 \cdot 20 = 8.4\ J/s$

A significant 40 % increase in heat flow rate. This shows that even a gentle flow around the object can speed up the cooling process significantly. More on that later.

If you read the first volume of this book, you know that we can easily calculate the amount of energy required to heat an object. To raise the temperature of an object from U to T, we need the energy:

$E = c \cdot m \cdot (T - U)$

with the mass of the object m and its specific heat capacity c (dependent on material). Since Newton's law of cooling enables us to calculate how much heat is transferred per second, we can estimate the cooling time by $t = E / P$ (in s).

$\mathbf{t \approx c \cdot m / (\alpha \cdot S)}$

One immediate result is that the larger the ratio of mass to surface area is, the longer the cooling process will take. Massive objects with a small surface area take a long time to cool while light objects with a large surface area cool quickly.

By the way: where did the temperature difference go? How can the cooling time not depend on the temperature difference? Well, both the excess energy and the heat flow rate are proportional to said difference. If an object has a higher temperature difference, it needs to give off more energy to be in thermal equilibrium with its environment. But at the same time, the rate at which the object gives off its heat is also increased. So the cooling time is not altered.

Let's see how long it takes the can of coke to cool in the fridge. The specific heat capacity is about $c = 3800\ J/(kg\ K)$ and the mass $m = 0.35\ kg$. For the surface area and heat transfer coefficient we'll use the values from the first example of this section: $S = 0.03\ m^2$ and $\alpha = 10\ W/(m^2\ K)$. Plugging it all into the formula for the cooling time results in:

$t \approx 3800 \cdot 0.35 / (10 \cdot 0.03)$

$t \approx 4430\ s \approx 74$ *minutes*

The cooling time is drastically reduced if we put the can into water. For stationary water the heat transfer coefficient is about $\alpha = 250\ W/(m^2\ K)$, which leads to the cooling time:

$t \approx 3800 \cdot 0.35 / (250 \cdot 0.03)$

$t \approx 177\ s \approx 3$ *minutes*

This demonstrates clearly how much better water is at conducting heat than air. Actually, air does such a bad job at transmitting heat that is considered to be a heat insulator.

Grandma sets her delicious m = 0.5 kg apple pie cake on the window board to cool. Assuming that the pie is r = 0.08 m in radius and h = 0.06 m in height, the surface area that is in contact with the surrounding air is about S = 0.05 m². From a table of specific heat capacities we find that the value c = 2000 J/(kg K) should be about right. Again we'll use α = 10 W/(m² K) for the heat transfer coefficient. How long will grandma's apple pie take to cool?

$t \approx 2000 \cdot 0.5 / (10 \cdot 0.05)$

$t \approx 2000 \ s \approx 33 \ minutes$

The examples show that the calculation itself is really easy. The hardest part is finding suitable values for the specific heat capacity and heat transfer coefficient. A site that can help here is engineeringtoolbox.com. It has a great collection of tables for many quantities commonly needed in physics.

There's one more thing I'd like to discuss before concluding this section and that is the dependence of the heat transfer coefficient on flow and the implications. In the second example we calculated that we can significantly increase the heat transfer by adding some movement in the surrounding air. Let's look at this from another angle.

Suppose it's U = 20 °C = 68 °F and no wind outside. Your body temperature is about T = 37 °C ≈ 100 °F. According to the Mosteller formula, a person with mass m (in kg) and height h (in cm) has a surface area of:

$S = \text{sqrt}(m \cdot h / 3600)$

For an average male adult ($m = 75$ kg and $h = 1.75$ m) we get the surface area $S \approx 1.91$ m². The fact that there's no wind means that the heat transfer coefficient is about $\alpha = 10$ W/(m² K). So this is the amount of heat your body would lose per second given these conditions:

$$P = 10 \cdot 1.91 \cdot 17 \approx 325 \text{ J/s}$$

This feels relatively comfortable. But now suppose the wind picks up. At a wind speed of 10 m/s ≈ 22 mph the heat transfer coefficient has a value of roughly $\alpha = 30$ W/(m² K). Accordingly, the heat flow rate increases to:

$$P = 30 \cdot 1.91 \cdot 17 \approx 974 \text{ J/s}$$

Despite the temperature not having changed a bit, it now feels much more chilly. So how warm we perceive it to be depends not only on the temperature, but also (and strongly) on the wind speed. The higher the wind speed, the higher the rate at which we lose heat to the environment.

Why is that? Well, each molecule that comes in contact with you picks up some of your heat (assuming the air is colder than your body, which it usually is). As the wind speed increases, so does the rate at which the molecules pass you. Hence, it makes a lot of sense to have the heat transfer coefficient depend on the speed of the surrounding material.

- **Adiabatic Processes:**

Usually when looking at gases, the focus lies on the ideal gas law. It is a simple and beautiful formula that connects the volume, pressure and temperature. Once you know all of these quantities in an initial state, you can do computations in any other state. For example, if the volume and pressure changed in a certain way, we can deduce the corresponding change in temperature from the ideal gas law. Or if the volume and the temperature were altered, we could predict how this affects the pressure.

So the ideal gas law is very broad. But this strength is at the same time its weakness. Often times we don't have the luxury of a having so much data at hand. For example we might know all the initial values and the change in volume, but have no additional information on pressure or temperature. Applying the ideal gas law is impossible then. Under certain circumstances however, there's a way out, a way, in which we can compute gases with only little data at hand. This neat trick can be accomplished with the formulas for adiabatic processes.

Under what circumstances can we apply these? The answer is very simple. We can apply the adiabatic formulas whenever the heat loss is negligible during the change of state. This is for example the case when the gas is in a well insulated container during the process. Another common example is when the process happens so quickly that no significant heat is lost (which is the case inside motor cylinders).

With this said, let's turn to the formulas. We need three quantities: the volume (any unit will do), the pressure (same here) and the temperature (in K). Note that we have to input the temperature in Kelvin rather than Celsius or Fahrenheit. Otherwise we'll end up with an incorrect result. Luckily the conversion from Celsius to Kelvin is rather simple, just add 273.15 to the temperature in Celsius to get the temperature in Kelvin. For Fahrenheit it's a little more complicated, but you can use the conversion formula in the appendix of this book or do it online.

During an adiabatic process (negligible heat loss) these expressions remain constant:

$$\mathbf{p \cdot V^{1.4} = const.}$$

$$\mathbf{T \cdot V^{0.4} = const.}$$

$$\mathbf{T / p^{0.3} = const.}$$

This is how we apply the formulas: we calculate the constants from an initial state, then we are free to compute any state we desire. To be pedantically precise I should note the above formulas are only valid for diatomic gases (such as air). For the rare case of dealing with monoatomic gases the exponents change to 1.7, 0.7 and 0.4 respectively.

Note that now only two quantities appear in each formula. This means that knowing the change in one variable is sufficient to determine how the other quantities change. This is what the ideal gas law did not allow us to do. The down side is that now we are limited to processes with negligible heat loss.

Let's turn to some examples.

In a motor cylinder the gas is constantly compressed and expanded. This happens at a very fast pace, so fast, that we don't need to worry about having significant heat losses during one cycle. Thus, we are free to apply the adiabatic formulas.

Assume that in the initial state the gas inside the cylinder has a volume $V = 2000\ cm^3$, temperature $T = 25\ °C$ (room temperature) and pressure $p = 1$ atm (air pressure at sea level). Now the gas is compressed by a factor of twenty to $V' = 100\ cm^3$. This compression ratio is common in diesel engines. How will this affect the temperature and pressure?

Let's look at the pressure first. From the initial state we can determine the constant. To keep things simple we won't use any units in the formulas (which you normally shouldn't do).

$p \cdot V^{1.4} = const.$

$1 \cdot 2000^{1.4} \approx 41{,}826$

Now that we know the constant, we can find out what happens to the pressure inside the cylinder.

$p' \cdot 100^{1.4} = 41{,}826$

$p' \cdot 631 \approx 41{,}826$

$p' \approx 66\ atm$

So the pressure increases by a factor of sixty-six. You certainly wouldn't want to be in the way of that. What about temperature? Again we take a look at the constant first. Don't forget to convert to Kelvin (25 °C are roughly 298 K).

$$T \cdot V^{0.4} = const.$$

$$298 \cdot 2000^{0.4} \approx 6232$$

Now we apply the formula one more time with the volume changed and the temperature being the unknown. We get:

$$T' \cdot 100^{0.4} = 6232$$

$$T' \cdot 6.3 \approx 6232$$

$$T' \approx 990\ K \approx 715\ °C \approx 1320\ °F$$

This is enough to ignite the gas inside the cylinder. As you saw, the formulas are extremely useful and not that hard to apply. It's only a few lines of calculations.

A parcel of air is heated by the hot ground below it and begins to rise. Initially its temperature is $T = 32$ °C and pressure $p = 1$ atm. As it rises, the pressure decreases. At 1 km altitude, the pressure has reduced to $p' = 0.9$ atm. How does this affect the temperature?

First of all: are we even allowed to use the adiabatic formulas here? The air parcel is neither in an insulated container nor is this process happening particularly fast. This is true, however, air is a very poor conductor of heat. So while not a container in the classic sense, the air surrounding the parcel does act as an insulator and thus heat loss is negligible.

Let's determine the constant using the initial values (again, don't forget the conversion to Kelvin, 32 °C = 305 K):

$T / p^{0.3} = const.$

$305 / 1^{0.3} = 305$

Now we can calculate the air temperature at 1 km height given the pressure change to 0.9 atm:

$T' / 0.9^{0.3} = 305$

$T' / 0.97 \approx 305$

$T' \approx 296\ K \approx 23\ °C$

So the air parcel cooled by roughly 9 °C while ascending.

It is worth noting that having ratios is sufficient for solving problems involving adiabatic processes. Remember that in the first example we had a quick look at compression in diesel engines. We were given the initial volume V = 2000 cm^3 (along with an initial temperature and pressure) and the final volume V' = 100 cm^3. However, stating that the gas is compressed by a factor of 20 would have been sufficient. In this case we could arbitrarily choose volumes by a factor of 20 apart and still arrive and the same final temperature and pressure. For example V = 20 and V' = 1 (no matter what unit) would work just fine.

In an airsoft gun the plastic pellets used as ammunition are accelerated via compressed air. When the trigger is pulled, a spring mechanism compresses the air on one side of the pellet, creating a pressure difference. This in turn leads to a force on the pellet. The resulting force is just the product of *the pressure difference Δp (in Pa) and the projected area A (in m^2) of the pellet:*

$$F = \Delta p \cdot A$$

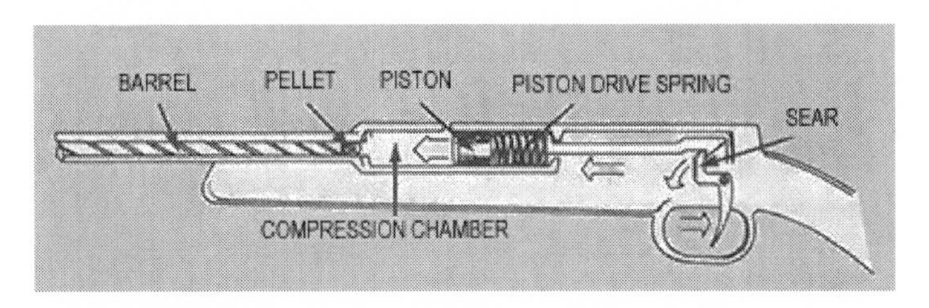

Assume that the air is compressed by a factor of 5. The initial pressure is $p = 1$ atm $\approx 100{,}000$ Pa. What is the resulting pressure? What force will the pellet experience if its projected area is $A = 0.00003\ m^2$ (a 6 mm pellet)?

To answer these questions, we need an initial and final volume. Since we are only given a compression ratio, we can choose suitable values. Let's go with the simplest possibility $V = 5$ and $V' = 1$. As always, we first determine the constant using the initial values:

$$p \cdot V^{1.4} = const.$$

$$100{,}000 \cdot 5^{1.4} \approx 952{,}000$$

Now we can easily deduce the final pressure:

$$p' \cdot 1^{1.4} = 952{,}000$$

$$p' = 952{,}000\ Pa$$

This of course means that the pressure difference is Δp = 952,000 - 100,000 = 852,000 Pa. The force is:

$$F = 852{,}000\ Pa \cdot 0.00003\ m^2 \approx 25.6\ N$$

This might not sound like a lot, but given the small mass of such a pellet (think Newton's second law: $a = F / m$), it translates into a significant acceleration. Note that as soon as the pellet starts moving, the volume of the compressed air rapidly increases and accordingly the pressure difference goes down.

While these formulas are indeed very useful, don't forget that perfect adiabatic conditions can never be achieved. There's bound to be some heat loss and the more there is, the more the theoretical values will differ from reality. Nevertheless, it's a great tool physicists wouldn't want to be without.

- **Draining a Tank**

Let's turn a problem often addressed when dealing with fluid flow, the process of draining a water-filled tank. We assume that there's no pump to speed up the process, all the outflow will solely be a result of the gravitational pull of the liquid. Some elementary questions that arise here are: at what speed and flow rate will the water leave the tank? How long does it take to drain the tank?

The key to solving this problem is Toricelli's law. It relates the outflow velocity v (in m/s) to the height h (in m) of the fluid above the opening. The only additional quantity involved here is the well-known gravitational acceleration g (in m/s^2).

v = sqrt (2 · g · h)

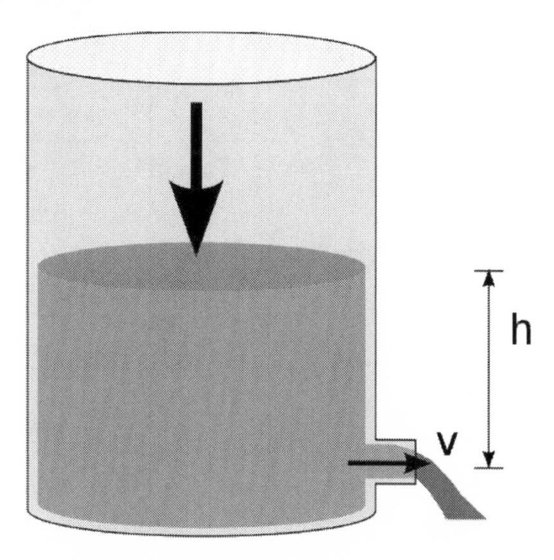

Note that this relationship implies that the outflow speed doubles when you quadruple the height of the liquid. Let's take this one step further. To calculate the time it takes to drain the tank, we need to know the volume flow rate r (in m^3/s). A quick look at the units will help. To get from speed (m/s) to flow rate (m^3/s), we obviously need to multiply the speed with an area (m^2). This way the units check out. So additionally to the speed, we must be given the area A (in m^2) of the opening.

$$\mathbf{r = \mu \cdot A \cdot v}$$

What about the μ in the formula? It is a dimensionless quantity called the discharge coefficient and depends in a complex way on the geometry of the opening. If the opening is rounded and smooth, μ will be close to one, meaning that the maximum theoretically possible volume flow rate is (almost) achieved. Any deviation from this optimal form will result in a lower discharge coefficient and thus lower flow rate. In the image below you can see some examples.

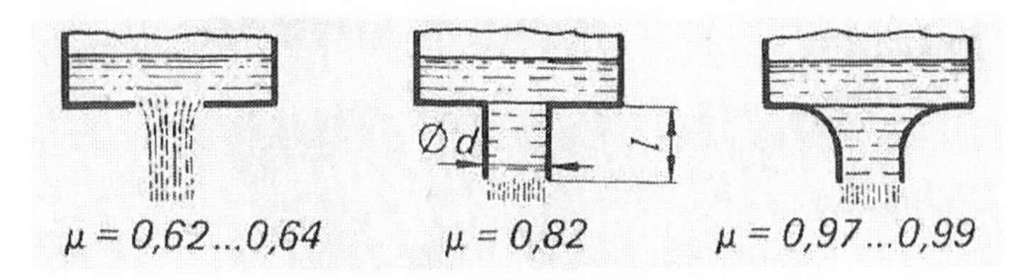

Before we go on to the draining time, let's look at an example to get familiar with the above formulas.

We fill a tank to a height of h = 1 m with water. Unbeknown to us, there's a small hole with an area of A = 0.0001 m^2 (circle with a radius of 6 mm) at the very bottom. Since it's just a common hole without any specific shape, we can assume from the above image that the discharge coefficient *is somewhere around* $\mu = 0.6$*. At what speed and flow rate will the water leave the tank?*

First let's compute the velocity using Toricelli's law. The gravitational acceleration at earth's surface is $g = 9.81\ m/s^2$*. Thus we get:*

$v = sqrt\ (2 \cdot 9.81\ m/s^2 \cdot 1\ m)$

$v \approx 4.4\ m/s \approx 16\ km/h \approx 10\ mph$

For the volume flow rate we get:

$r = 0.6 \cdot 0.0001\ m^2 \cdot 4.4\ m/s$

$r \approx 0.00026\ m^3/s \approx 0.95\ m^3/h$

Now it might be tempting to say that if the tank initially contains $9.5\ m^3$ *of water, it should take* $t = 9.5\ m^3 / 0.95\ m^3/h = 10$ *hours for the tank to empty. Unfortunately, it's not that simple. Why? As the water flows out, the water height decreases and with it the outflow speed and flow rate. So both quantities are variable throughout the process and doing the ratio using the initial volume flow rate is not sufficient.*

However, the above ratio is still useful as it provides a lower bound for the time it takes to empty the tank. It will not take 10 hours, but whatever the correct duration, it must be more than that.

Deriving the formula for the outflow time requires some calculus, so we will look at it only briefly. Denoting the (constant) cross-sectional area of the tank by T (in m²), this relationship must hold true:

$T \cdot dh/dt = - \mu \cdot A \cdot sqrt(2 \cdot g \cdot h)$

On the left side is the rate of change of the water volume inside the tank, on the right side the formula for the volume flow rate through the opening with Toricelli's law inserted. After separating the variables and integrating both sides, we get this expression for the time t (in s) it takes the water level to reach the height of the opening:

$\mathbf{t = 2 \cdot T / (\mu \cdot A) \cdot sqrt(h / (2 \cdot g))}$

Excuse the bracket orgy, unfortunately Kindle does not let me format this in a more humane way. Note that here h refers to the initial height of the water above the opening. Now let's revisit the previous example to apply this fantastic formula.

We return to the tank initially filled to $h = 1$ m. As stated, the area of the opening is $A = 0.0001$ m^2 and the discharge coefficient $\mu = 0.6$. We assume the cross-section of the tank to be $T = 9.5$ m^2 (the total volume is then 9.5 m^3 as before). How long will it take until the tank is empty? Remember that we already have a lower bound of 10 hours.

Let's apply the formula:

$t = 2 \cdot 9.5 / (0.6 \cdot 0.0001) \cdot sqrt(1 / (2 \cdot 9.81))$

$t = 19 / 0.00006 \cdot sqrt(1 / 19.62)$

$t = 316{,}666 \cdot sqrt(0.051)$

$t \approx 71{,}500\ s \approx 20\ h$

Don't be intimidated by the formula. If you work the result out step by step as I did here, nothing will go wrong.

We fill our rectangular bathtub (2 m by 1 m, so $T = 2\ m^2$) to a height of $h = 0.3$ m with water. The area of the opening is $A = 0.005\ m^2$ (circle with a radius of 4 cm) and we assume $\mu = 0.8$. Calculate the initial outflow velocity and volume flow rate as well as the time it takes for the bathtub to empty.

First let's look at the outflow speed:

$v = sqrt\ (2 \cdot 9.81\ m/s^2 \cdot 0.3\ m) \approx 2.4\ m/s$

Now onto the initial volume flow rate:

$r = 0.8 \cdot 0.005\ m^2 \cdot 2.4\ m/s \approx 0.0096\ m^3/s$

And here's the outflow time:

$t = 4 / 0.004 \cdot sqrt\ (0.3 / 19.62)$

$t \approx 124\ s \approx 2\ min$

That was relatively painless. By the way: how would these quantities change if we optimized the geometry of opening, bringing it (close) to $\mu = 1$? The outflow speed would remain the same, however the initial volume flow rate would change to $r = 0.012\ m^3/s$ (25 % increase) and the time it takes to drain the bathtub to $t = 100$ s (20 % decrease).

Finally a word on how changing the inputs affects the outflow time. We'll look at each variable in the formula.

- If you double the cross-sectional area T of the tank, the outflow time doubles as well. This makes sense since (with the height of the water unchanged) the tank now contains much more water.

- If you double the area A of the opening, the outflow time halves. So we could drain the above bathtub in about a minute if we increased the area of the opening to $A = 0.01 \text{ m}^2$ (circle with a radius of about 5.5 cm).

- If you quadruple the initial height h of the water, the outflow time doubles. A linear relationship might have been more satisfying for your (and my) intuition, but physics has its own mind.

- Last but not least, a lower gravitational acceleration g results in a higher outflow time (which makes sense since the outflow is driven by gravity alone). On the moon our (non-optimized) bathtub would take $304 \text{ s} \approx 5 \text{ min}$ to empty.

Note that all of this is a result of rigorously and stubbornly applying Toricelli's law. It is amazing how much you can extract from such a simple law with the right ideas (looking at the units) and mathematical tools (calculus).

- **Open-Channel Flow**

There's one thing most physics books, often even those focusing on fluid flow, have in common: they tend to ignore open-channel flow. This term refers to liquid flow with a free surface, such as the flow of water in creeks and lakes. However, we will go against the tide and take a detailed look at this very interesting and colorful topic.

Our starting point will be the Gauckler-Manning formula. Its history goes back to 1867 when the French engineer Philippe Gauckler first published it. The formula is used to compute the average flow speed in open-channel flow systems that are driven solely by gravity (which is the case for almost all creeks and lakes).

What does the average speed of the flow depend on? One important factor is the slope of the water surface. Suppose you move a horizontal distance Δx and the surface of the water drops by a height of Δh. The corresponding slope S (dimensionless) is:

$$S = \Delta h / \Delta x = \tan \theta$$

Alternatively and as indicated in the formula above, if we know the angle θ at which the surface is to the horizontal, we can just apply the tangent to the angle to get the slope. What other quantities impact the flow speed? There are two more to take into consideration.

Additionally to the slope we need to know the cross sectional area A (in m^2) and the wetted perimeter P (in m). The latter is the circumference of the the cross sectional area excluding the free surface. Engineers like to combine these two quantities into a one, the hydraulic radius R (in m):

$R = A / P$

Don't let the word "radius" fool you, we are not necessarily dealing with any circular shapes here. Hydraulic radius is just the term given to the above ratio, independent of the shape of the flow system. Determining this quantity accurately can be very problematic since natural flow systems don't stick to our idealized geometric shapes.

Despite this complexity, let's take a look channels with a rectangular cross section. Often times this approximation to reality will be sufficient. We'll denote the width of the channel by w (in m) and its depth by d (in m). The cross sectional area is then $A = w \cdot d$ and the wetted perimeter $P = w + 2 \cdot d$, leading to this formula for the hydraulic radius:

$R = w \cdot d / (w + 2 \cdot d)$

With the quantities so neatly defined and everything set in place to compute them, we can now easily state and apply the Gauckler-Manning formula. The average velocity v (in m/s) in a gravity-driven, open-channel flow system is:

$\mathbf{v = k \cdot R^{0.67} \cdot sqrt(S)}$

The factor k (in $m^{0.33}/s$) is called the Strickler coefficient and depends on the roughness of the ground. For rough mountain torrents k is around 10-20, for common rivers around 20-40 and can go as high as 100 for smooth concrete surfaces.

A very useful addition to the Gauckler-Manning formula is the equation for computing the volume flow rate Q (in m^3/s).

$\mathbf{Q = A \cdot v}$

You know the drill: now come the examples.

The Rhine river covers a horizontal distance of about 300 km while flowing from Cologne to the Northern Sea. Over this distance it drops about 50 m in height. For the most parts its width is around w = 350 m and depth d = 8 m. We set the Strickler coefficient to k = 30 $m^{0.33}$/s (a value which is supported by measurements). Determine the average flow speed for the Rhine river.

First we need to compute the slope. Make sure to input both distances in the same unit (that is, don't mix m and km).

$S = 50\ m / 300{,}000\ m \approx 0.00017$

For lack of more detailed information, we assume a rectangular cross section. This leads to the following value for the hydraulic radius:

$R = 350\ m \cdot 8\ m / 366\ m \approx 7.65\ m$

Now we can apply the Gauckler-Manning formula to determine the average flow speed (I'll leave out the units):

$v = 30 \cdot 7.65^{0.67} \cdot sqrt(0.00017)$

$v \approx 1.5\ m/s \approx 5.5\ km/h \approx 3.4\ mph$

This value agrees quite well with observations, a solid victory for Gauckler and Manning despite our crude assumption of a rectangular shaped cross section.

We estimate the flow speed of a small mountain creek to be $v = 2$ m/s. Its width and depth is $w = 1.2$ m and $d = 0.2$ respectively. What is the slope of the mountain? What is the volume flow rate? Assume a rectangular cross section and use the value $k = 15\ m^{0.33}/s$ for the Strickler coefficient.

In a first step we'll determine the hydraulic radius. Applying the formula for rectangular shaped channels we get:

$$R = 1.2\ m \cdot 0.2\ m / 1.6\ m = 0.15\ m$$

With the Gauckler-Manning formula and all the given data we set up an equation for the slope:

$$2 = 15 \cdot 0.15^{0.67} \cdot sqrt(S)$$

$$2 \approx 4.21 \cdot sqrt(S)$$

Divide by 4.21:

$$0.48 \approx sqrt(S)$$

Finally square both sides:

$$S \approx 0.23$$

So over a horizontal distance of 100 m the height will drop by 23 m. We can also express this in form of an angle:

$$\theta \approx arctan(0.23) \approx 13°$$

Onto the volume flow rate. The cross sectional area is $A = 1.2\ m \cdot 0.2\ m = 0.24\ m^2$. So we get the volume flow rate:

$$Q = 0.24\ m^2 \cdot 2\ m/s = 0.48\ m^3/s \approx 1730\ m^3/h$$

We'll continue the topic of open-channel flow with a structure that is commonly found in such systems: the weir. A weir is an obstruction in the channel that allows the water to flow over it. They are put in place to prevent flooding and to help determine the volume flow rate of a creek or river. If done properly, the discharge can be measured with an accuracy of about ± 5 %.

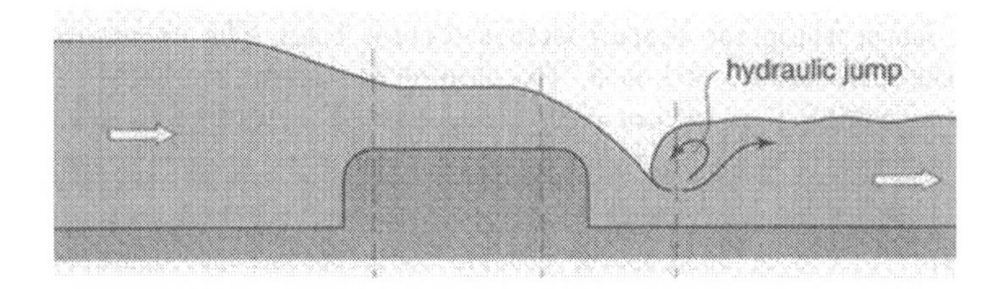

To compute the volume flow rate Q (in m^3/s) over the weir, which is identical to the flow rate in the river, we need to know the height h (in m) of the water above the weir and the width w (in m) of the weir.

$$Q = C \cdot w \cdot h^{1.5}$$

The so called discharge coefficient C depends on the type of weir. For broad-crested weirs such as the one shown in the above image C is around 2.5 while for sharp-crested weirs this increases to about 3.2. Of course the formula $Q = A \cdot v$ is still valid, so we can easily derive an expression for the average flow speed v (in m/s) over the weir. Inserting $A = b \cdot h$ and solving for the velocity results in:

$$v = C \cdot sqrt(h)$$

Let's apply these formulas.

We build a broad-crested weir ($C = 2.5$) with the intention of measuring the volume flow rate in a small river of width $w = 6$ m. After the construction is done, the height of the water above the weir is measured to be $h = 0.2$ m. Compute the volume flow rate.

Applying the formula we get:

$$Q = 2.5 \cdot 6 \text{ m} \cdot (0.2 \text{ m})^{1.5} \approx 1.34 \text{ m}^3/\text{s} \approx 4820 \text{ m}^3/\text{h}$$

Since at best we should expect an accuracy of ± 5 %, it's wise to express this result in the following form:

$$Q \approx 4820 \pm 240 \text{ m}^3/\text{h}$$

From applying the Gauckler-Manning formula we know that the water approaches a certain weir at the speed $v = 0.5$ m/s. The river is $w = 8$ m wide and $d = 0.6$ m deep. The broad-crested weir ($C = 2.5$) extends over the entire width of the river. Assuming the river has a rectangular cross section, determine the height of the water above the weir and the corresponding flow speed.

From the downstream velocity of the river and its dimensions we can calculate the volume flow rate:

$$Q = A \cdot v = 8 \text{ m} \cdot 0.6 \text{ m} \cdot 0.5 \text{ m/s} = 2.4 \text{ m}^3/\text{s}$$

With this done, we can set up an equation for the height above the weir using the weir formula:

$$2.4 = 2.5 \cdot 8 \cdot h^{1.5}$$

$2.4 = 20 \cdot h^{1.5}$

Divide by 20:

$0.12 = h^{1.5}$

Now take both sides to the power of 2/3 ≈ 0.67 (which is the reciprocal to 3/2 = 1.5):

$h \approx 0.12^{0.67} \approx 0.24\ m$

The flow speed over the weir can now easily be determined:

$v = 2.5 \cdot sqrt(0.24) \approx 1.22\ m/s$

So the weir causes the water to speed up by a factor of about 2.5. This is what can make weirs dangerous to swimmers. The increased speed can cause them to lose control and the rough water conditions just behind the weir can submerge a person.

For the next section we'll leave the creeks and rivers behind us have a look at oceans or rather the wind-driven waves within them. Again, this is something most physics books do not cover, but I can assure you that it's a very interesting topic.

- **Wind-Driven Waves**

I admit that this formula, or rather the search for it, almost drove me crazy. Since the formula is very little known outside a few niches and in papers often just referenced rather than reproduced, it took quite a while to find reliable information on it and cost me at least one sleepless night. But luckily the search was not in vain and with some pride I present to you another great formula called Greager's formula (which is a modification of an equation called Stephenson's formula).

The title of this section already gives you an idea what the formula is about. It allows us to estimate the height of wind-driven waves. What are the factors that determine wave height? One factor is time span over which the wind blows. If the wind is strong but only blows for a short period of time (less than an hour), the water is not able to absorb the energy and accordingly you won't see any significant waves building up. For the rest of this section when I talk about wind, I refer to a steady wind that has been blowing long enough to cause the water to react (several hours).

What other factors are there? The obvious factor is the wind speed. It is what leads the water surface to experience stress and to absorb the energy contained in the wind. Both the stress and energy grow approximately quadratically with wind speed. However, duration and wind speed alone don't do the trick.

Additionally to that, we need to have a large fetch. This term comes from the fields of geography and meteorology and refers to the length over which the wind blows. It is the final ingredient in producing powerful wind-driven waves.

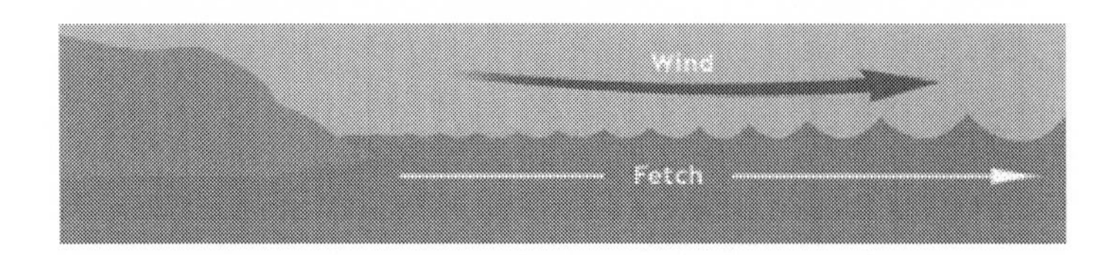

Greager's formula is an empiric formula that allows the estimation of wave height h (in ft), measured from trough to crest, from the wind speed v (in mph) and the fetch F (in miles). The adjective "empirical" means that the formula is not derived from theoretical considerations but rather came to be as a fit to real-world data.

$$h = 0.29 \cdot sqrt(v) \cdot F^{0.35}$$

One limitation is that it only holds true if the fetch is larger than 20 miles. Another interesting quantity in this context is the energy density E of the waves. It tells us how much energy the waves contain per square meter and thus is measured in the unit J/m^2. It can be estimated from the wave height h (in ft) using this formula:

$$E = 56 \cdot h^2 = 4.7 \cdot v \cdot F^{0.7}$$

So an immediate result of Greager's formula is that the energy content of wind-driven waves grows linearly with the wind speed. As a final addition to our discussion, I'll include one more formula. It was derived by Ekman and allows us to compute the velocity of the water waves w (in mph) from the wind speed v (in mph):

$$w = 0.02 \cdot v$$

So there's a neat linear relation between water wave and wind velocity without the fetch being an influencing factor. An interesting side note: the water waves generated by the wind don't travel in the direction the wind blows. The angle between the two is always around 45°. The reason for that is the rotation of earth. It results in a force deflecting the water waves that otherwise would travel in unison with the wind.

A steady wind blows for several hours from the Bahamas to Bermuda ($F = 900$ miles) at a speed of $v = 30$ mph. What wave heights can we expect? What's the corresponding energy density of the waves? And how fast do the waves travel?

The answer to the first question we can estimate using Greager's formula. We get:

$h = 0.29 \cdot sqrt(30) \cdot 900^{0.35}$

$h \approx 17 ft \approx 5 m$

The energy density is:

$E = 56 \cdot 17^2 \approx 16{,}200 J/m^2$

Last but not least we'll take a look at the speed at which the generated water waves travel:

$w = 0.02 \cdot 30 mph = 0.6 mph = 53 ft/min$

The wind speed changes to $v = 50$ mph, while the fetch remains the same. How does this affect the sea conditions? First let's look at the wave height:

$h \approx 0.29 \cdot sqrt(50) \cdot 900^{0.35}$

$h \approx 22\ ft \approx 6.7\ m$

And the energy content:

$E = 56 \cdot 22^2 \approx 27{,}100\ J/m^2$

Finally the speed of the waves:

$w = 0.02 \cdot 50\ mph = 1\ mph = 88\ ft/min$

As the waves come in, we observe them having a height of $h = 12\ ft$ and velocity of $w = 0.5\ mph$. What can we say from that about the wind system that generated them? It turns out a lot.

From the speed of the waves we can estimate the speed of the wind that generated them.

$0.5\ mph = 0.02 \cdot v$

Divide by 0.02:

$v = 25\ mph$

Now that we have an estimate for the wind speed, we can use Greager's formula to find out the fetch:

$12 = 0.29 \cdot sqrt(25) \cdot F^{0.35}$

$12 = 1.45 \cdot F^{0.35}$

Divide by 1.45:

$8.3 \approx F^{0.35}$

Taking both sides to the power of 2.86 (which is the reciprocal of the exponent 0.35) leads to:

$F \approx 425$ *miles*

So we can find out quite a bit by reversing the situation. Note that for the results to be realistic, we shouldn't observe the waves too close to the shoreline. This is because in shallow water the ground interacts with the waves, causing them to slow down and pile up.

Never underestimate the power of empirical equations. Despite them not being rooted in theoretical considerations and often quite limited in the range in which they can be applied, they provide a great way to get a quick estimate for complex situations.

- **Sailing**

Long before motors came along, people were able to cross the seas by making use of wind power. Sailing has been mankind's number one tool for exploring the world for centuries. It is also quite interesting from a physical point of view. Sailing provides a great example of how to use the concept of force equilibrium to derive an easy-to-apply formula.

One note before we start: we will only talk about classic sailboats such as the ones used by the great explorers of medieval times. The physics behind modern sailboats is much more complex and beyond the scope of the book. But rest assured, you can learn a lot about physics from the classic sailboats as well. So let's get started.

One interesting aspect of sailing is that both the driving and the retarding force have the same nature: they are a result of drag from motion relative to a fluid. The driving force is the drag from air flowing past the sail, the retarding force the drag from water flowing past the hull.

Note however, and this is a crucial point, that the relative speed is not the same for these two cases. Let's assume the water to be stationary. If the wind speed is w (in m/s) and the boat moves in the direction of the wind at the velocity v (in m/s) relative to the water, then the relative speed between the air and the sail is w - v. Keep that in mind.

For both forces we will use the same formula. It is the formula also used for computing the terminal velocity in free fall. If an object moves at a speed of v relative to a fluid of density D (in kg/m^3), then the drag F (in N) is:

$F = 0.5 \cdot c \cdot D \cdot A \cdot v^2$

with the area relative to the fluid A (in m^2, also called the projected area) and the drag coefficient c. You already saw this formula in the section "Maximum Car Speed".

Let's look at the force pushing the sailboat. From a table of drag coefficients we can see that a value $c = 1.4$ seems about right for classic sails. The density of air at sea level is $D = 1.25\ kg/m^3$. Denoting the the area of the sail by S and remembering that the relative speed between the air and sail is $w - v$, we get this formula for the driving force:

$F(d) = 0.88 \cdot S \cdot (w - v)^2$

Onto the retarding force. Sailboats are streamlined, so the drag coefficient should be around $c = 0.04$. The density of water is approximately $D = 1000\ kg/m^3$. With the projected area A, which is determined by the draft of the boat, we get this expression for the retarding force:

$F(r) = 40 \cdot A \cdot v^2$

Now the concept of force equilibrium comes into play. After some time the boat will reach a certain equilibrium speed (this is of course assuming constant wind speed, which we do). For a constant speed to be maintained, the forces must cancel each other. So from the equation $F(d) = F(r)$, we can derive this great formula for the equilibrium speed of the sailboat:

$v = w \cdot sqrt(S/A) / (4.8 + sqrt(S/A))$

Note that for a small sailboat the draft is about 1.2 m, leading to $A = 1.4\ m^2$. For very large sailboats such as the Rainbow Warrior (owned by Greenpeace) the draft can be as much as 5 m and the corresponding projected area A = 25 m^2.

One immediate and simple result: the equilibrium speed of the boat is proportional to the wind speed. If the wind speed doubles, so does the speed of the boat. The dependence on the area of the sail is not so simple. For smaller sails it grows with the square of the area, so quadrupling the area leads to a doubling in speed. For larger sails the dependence gets weaker. As the area of the sail grows to infinity (mathematicians love having a quantity grow to infinity, don't worry about it), the boat speed approaches the wind speed.

Assume we are on a small sailboat ($A = 1.4\ m^2$) and the current wind speed is $w = 10$ m/s ≈ 22 mph. How fast can we expect to go if the area of the sail is $S = 40\ m^2$? How long would a 100 mile journey take under these conditions?

Let's use the formula:

$v = 10 \cdot sqrt(28.6) / (4.8 + sqrt(28.6))$

$v \approx 53.5 / 10.1 \approx 5.3$ *m/s ≈ 11.8 mph*

So the journey would take:

$t = 100$ *miles / 11.8 mph ≈ 8.5 h*

This is way too long for our taste. So we switch to a $S = 70\ m^2$ sail. What effect does this have on the speed of the boat and the time it takes to complete the journey?

$v = 10 \cdot sqrt (50) / (4.8 + sqrt (50))$

$v \approx 70.7 / 11.9 \approx 5.9$ m/s ≈ 13.3 mph

So the journey would take:

$t = 100$ miles / 13.3 mph ≈ 7.5 h

We save roughly on hour this way.

Let's take a look at the Rainbow Warrior. As stated, it has a projected area of about $A = 25\ m^2$. The sail area is $S = 1255\ m^2$. What ratio of boat to wind speed can it achieve? From the formula we get:

$v / w = sqrt (50.2) / (4.8 + sqrt (50.2))$

$v / w \approx 0.6$

So it should be able to go to 60 % of the current wind speed.

After dedicating the last few sections to all things water related, I think it's time to dry ourselves and move on. Though there will be flow in the next section as well, it won't involve any liquids or gases.

- **Heat Radiation**

Heat can be transported in three ways: by conduction, by convection and by radiation. In this section we will focus on the latter, but despite that, let's quickly go over all the mechanisms to get the full picture.

Temperature is associated with the vibrational and kinetic energy of molecules. The hotter an object is, the stronger the molecules within it vibrate (solids) or move (gases). If two objects come into direct contact, their molecules collide and energy is transferred. The molecules of the hotter object lose some of their energy to the molecules of the cooler object. This process continues until the energies of the molecules are equalized. The objects have now reached the same temperature. This process of heat flow on a molecular level is called heat conduction.

Heat convection involves a fluid transporting heat. For example, consider a parcel of air that picks up heat by direct contact with the ground. When it rises, it gives off this heat to the air located at a higher altitude. Again, heat has been transported, but on a larger scale. Instead of going from molecule to molecule, the air parcel picked up the heat and delivered it to another location.

Now think of the sun. We can certainly feel its heat arriving on earth, but how does it get here? Because of the vacuum between the sun and earth, conduction or convection is not possible. There are no molecules in contact and there's no fluid that could pick up and deliver the heat. Rather, the energy gets here by means of electromagnetic radiation: radio waves, microwaves, infrared light, visible light, and so on. You can think of this radiation as a stream of photons all having a certain frequency and thus energy. When a photon emitted by the sun reaches earth, it is absorbed and earth becomes the proud new owner of the energy it carried. This is how heat radiation works.

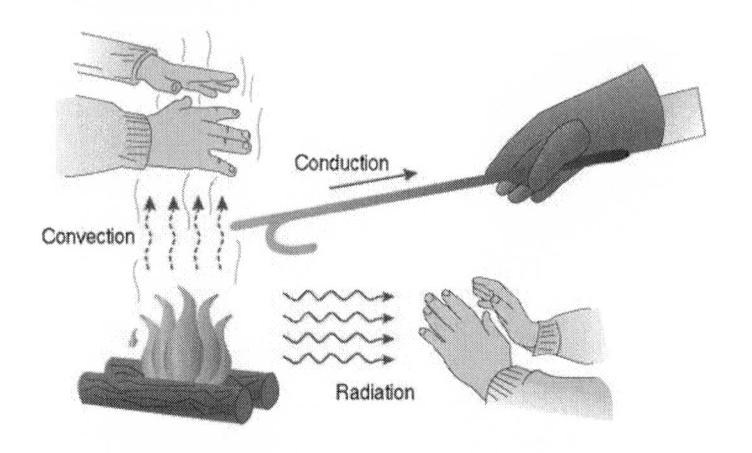

Let's turn to the mathematical description by thinking about what quantities determine how much heat an object gives off by radiation. One obvious candidate is the temperature. The hotter an object is, the more heat it should give off in any second. Less obvious, but just as important is the surface area, because the greater the area of contact is, the more intense the heat exchange should be.

There's a beautiful formula that allows us to compute the heat radiation power P (in J/s = W) from the temperature T (in K) and the surface area A (in m^2). It's called the Stefan-Boltzmann law and looks like this:

$$\mathbf{P = \varepsilon \cdot \sigma \cdot A \cdot T^4}$$

σ is the Stefan-Boltzmann constant and has the fixed value $\sigma = 5.76 \cdot 10^{-8}$. So nothing to worry about. ε is called the emissivity and depends on the material and surface condition of the radiator. Black and dull objects have an emissivity close to one (it can't go past that), while reflective objects can have emissivities close to zero.

Note the extremely strong dependence of radiation power on temperature. If you double the temperature of the radiator, the power increases sixteen-fold. Such a strong variation is very rare in physics. The dependence on surface area on the other hand is just linear, so doubling the surface area will also double the radiation power.

Common heat radiators in households have temperatures around $T = 70\ °C = 343\ K$. What is the heating power of the radiator with a surface area of $A = 1\ m^2$ and an emissivity of $\varepsilon = 0.9$? We can answer this by directly applying the Stefan-Boltzmann law:

$$P = 0.9 \cdot 5.76 \cdot 10^{-8} \cdot 1\ m^2 \cdot (343\ K)^4$$

$$P \approx 718\ W$$

Let's put this number into perspective. Raising the temperature of air in a 5 m by 3 m by 3 m room by 5 °C requires approximately this amount of energy (see section "Heat" in volume I of this book):

$$E = 1000\ J/(kg\ °C) \cdot 1.25\ kg/m^3 \cdot 45\ m^3 \cdot 5\ °C$$

$$E = 281{,}250\ J$$

Assuming no heat is lost to the surrounding or any other objects in the room, our radiator could accomplish this in:

$$t = E / P \approx 392\ s \approx 6 \text{ and } 1/2 \text{ minutes}$$

The actual time is going to be a little lower than that since there's also a fair amount of convection taking place (if you put your hand over the radiator you can feel the hot air rising).

Just recently I had the opportunity to be around a PFR lime kiln, which is used to transform the limestone mined in a quarry to lime. This lime is further processed to create cement. The heat radiation coming off it was so intense that you were hardly able to keep your eyes open.

Assuming the temperature of the kiln's outside surface to be around $T = 350\ °C = 623\ K$*, how much heat does it radiate per unit area? We assume the emissivity to be* $\varepsilon = 0.8$*.*

$$P/A = 0.8 \cdot 5.76 \cdot 10^{-8} \cdot (623\ K)^4$$

$$P/A \approx 7000\ W/m^2$$

What about the inside surface? Inside the kiln the temperature is around $T = 1200\ °C = 1473\ K$*. Assuming the same emissivity, we get:*

$$P/A = 0.8 \cdot 5.76 \cdot 10^{-8} \cdot (1473\ K)^4$$

$$P/A \approx 217{,}000\ W/m^2$$

With this power you could heat the 5 m by 3 m by 3 m room from the first example in a little over a second.

In the first volume of this book we were able to conclude that the power output of the sun is: $P \approx 4 \cdot 10^{26}\ W$*. By* *observation the radius of the sun can be determined to be: R ≈ 695,500 km. What is its surface temperature? We'll set* $\varepsilon = 1$.

First we need to compute the surface area. We can use the corresponding formula for spheres (remember to input the radius in meters, not kilometers):

$$A = 4 \cdot \pi \cdot R^2 \approx 6.08 \cdot 10^{18}\ m^2$$

Now we can set up an equation for sun's surface temperature using the Stefan-Boltzmann law:

$$4 \cdot 10^{26} = 5.76 \cdot 10^{-8} \cdot 6.08 \cdot 10^{18} \cdot T^4$$

$$4 \cdot 10^{26} = 3.5 \cdot 10^{11} \cdot T^4$$

Divide by $3.5 \cdot 10^{11}$:

$$T^4 \approx 1.14 \cdot 10^{15}$$

Take the fourth root (which is the same as raising both sides to the power of one-fourth):

$$T \approx 5{,}800\ K$$

This result agrees quite well with observations.

Let's take this a step further. We can use the Stefan-Boltzmann law to derive a general formula for the average temperature of a planet. A distance x (in m) from a star with the power output P (in W), the intensity I (in W/m^2) of the star's heat radiation will be:

$$I = P / (4 \cdot \pi \cdot x^2)$$

Assume this radiation hits a planet with radius r (in m). Its projected area in direction of the rays is: $A = \pi \cdot r^2$. Thus it will absorb this amount of heat per second:

$$P' = I \cdot \pi \cdot r^2 = P \cdot r^2 / (4 \cdot x^2)$$

According to the Stefan-Boltzmann law, the planet will emit this amount of heat per second given its temperature T (in K):

$$P'' = \varepsilon \cdot \sigma \cdot 4 \cdot \pi \cdot r^2 \cdot T^4$$

Note that for the planet's surface area we again used the corresponding formula for spheres. At the equilibrium temperature, the heat absorbed must equal the heat given off.

$$P' = P''$$

All that's left is to solve for the temperature. The result is the general expression of the equilibrium temperature of a planet.

$$\mathbf{T^4 = P / (\varepsilon \cdot \sigma \cdot 16 \cdot \pi \cdot x^2)}$$

Did you notice what happened to the radius of the planet? Since both the heat absorption as well the heat emission is proportional to the square of a planet's radius, it simply canceled out. So the equilibrium temperature does not depend on the size of the planet.

Let's look at an example before discussion the dependencies.

Mars is at a distance of about $x = 2.3 \cdot 10^{11}$ m from the sun. As stated before, the power output of the sun is $P \approx 4 \cdot 10^{26}$

W. For lack of more reliable data, we'll assume the emissivity to be $\varepsilon = 0.9$ *(sand / dry soil). What is the surface temperature of Mars?*

Let's apply the formula. For more clarity we'll first compute the value of the denominator and leave out the units:

$\varepsilon \cdot \sigma \cdot 16 \cdot \pi \cdot x^2 =$

$0.9 \cdot 5.76 \cdot 10^{-8} \cdot 16 \cdot \pi \cdot (2.3 \cdot 10^{11})^2 \approx 1.4 \cdot 10^{17}$

Onto the formula:

$T^4 \approx 4 \cdot 10^{26} / 1.4 \cdot 10^{17} \approx 2.9 \cdot 10^9$

$T \approx 231\ K \approx -42\ °C$

Which agrees fairly well with the average of about -50 to -55 °C recorded by Mars rovers.

Now let's draw some conclusions from the equation for the planetary temperature:

- The temperature is proportional to the fourth root of the star's power output. This implies that if the power of the star increases sixteen-fold, the temperature of the planet doubles.

- The planetary temperature is inversely proportional to the square of the distance to the star. So if the distance quadruples, the temperature of the planet halves.

- As for the emissivity, if it increases (and with it the planet's heat emission), the planet's temperature decreases. For example, if Mars were covered in ice, its emissivity would grow to $\varepsilon = 0.97$. This would lower the equilibrium surface temperature by about 3 °C. Interestingly enough, if instead the surface were covered in snow ($\varepsilon = 0.8$), the temperature would actually grow by roughly 8 °C.

Note that the formula and the above conclusions only hold true if the planet has no or only a very thin atmosphere. I hope you liked this short bit on astronomy, because in the next section we will stick to this topic by looking at main sequence stars.

- **Main Sequence Stars**

With the advance of technology over the last decades, scientists were able to gather more and more reliable information on the luminosity and surface temperature of distant stars. If you plot these measurements in a luminosity - temperature coordinate system, you can make an interesting discovery. Most of the stars lie on a specific line through the coordinate system (see image below).

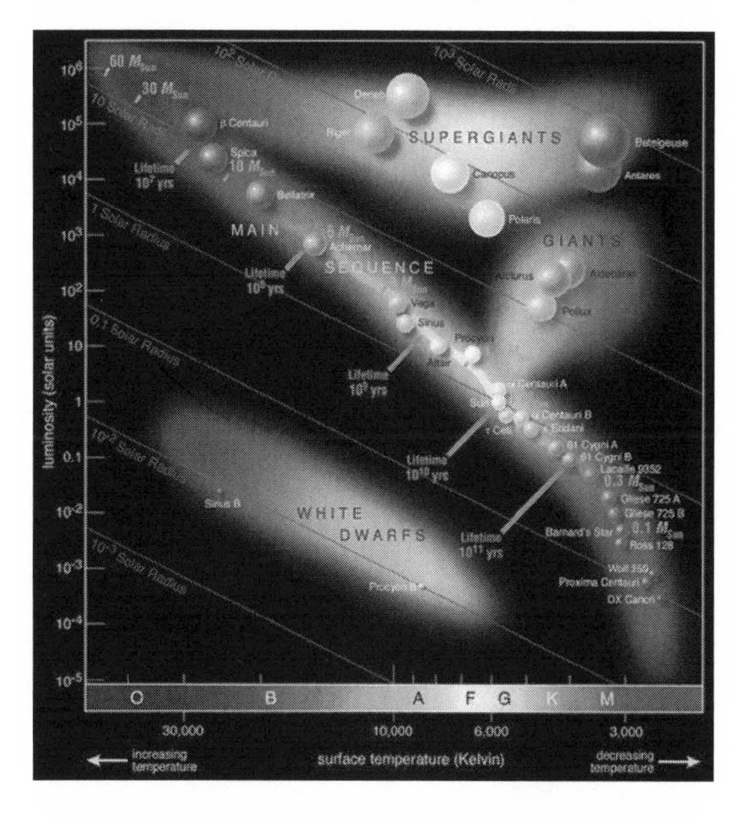

This line is called the main sequence and the stars on it (not surprisingly) main sequence stars. The main sequence thus provides us with a standard model of stars. It tells what luminosity a star with a certain temperature should have and vice versa. And it enables us to do many simple order-of-magnitude calculations on such stars. Here are some equations we can use for that.

Let's first take a look at the power output P (in W). Observations show that for main sequence stars the power output increases with the cube of the star's mass M (in kg):

$$\mathbf{P' / P = (M' / M)^3}$$

To do calculations with this formula, we need reference values for the power and mass. Luckily we know most of the parameters quite well for our sun, so we will always use it as the point of reference. Let's take a quick look on how to use this formula before expanding on it.

The brightest star in our night sky is Sirius (Alpha Canis Majoris). Its mass M' is about twice the mass of the sun M:

$$M' / M = 2$$

How does affect the of power output? We can get a rough estimate using the formula above:

$$P' / P = (M' / M)^3 = 2^3 = 8$$

So the power of Sirius should be about 8-times that of the sun. As you can see, the formula, as well as all the following, are very easy to apply.

A star's energy content, which is determined by the amount of burnable fuel it possesses, is approximately proportional to its total mass:

$E' / E = M' / M$

If we are given the amount of energy E and we are using it at the rate P, the time it takes for the energy to be depleted is $T = E / P$. Since for stars E is proportional to the mass and P proportional to the cube of the mass, the life span will thus be inversely proportional to the square of the mass:

$\mathbf{T' / T = (M / M')^2}$

So bigger stars have a shorter life.

Let's calculate the life span of Sirius in relation to our sun. Again we will denote the parameters of Sirius with primed symbols.

With $M' / M = 2$ (and thus $M / M' = 0.5$) we get:

$T' / T = (M / M')^2 = 0.5^2 = 0.25$

So Sirius will only live 1/4 as long as the sun. Scientists estimate the life span of the sun to be about 10 billion years, which means that the corresponding value for Sirius is about 2.5 billion years.

Another quantity we might be interested in knowing (and which will prove to be useful in drawing further conclusions) is the radius R (in m) of a star. For main sequence stars the radius scales with the mass to the power of 0.6:

$$\mathbf{R' / R = (M' / M)^{0.6}}$$

This relationship means that when the mass triples, the radius doubles. We will use this result two derive two other formulas. In the previous chapter you learned that the power output of a radiator is proportional to the surface area (which itself is proportional to the square of the radius) and to the temperature to the fourth power:

$$P \sim R^2 \cdot T^4$$

This is just the short form of the Stefan-Boltzmann law. We used the symbol ~ to indicate proportionality. Let's take both sides to the power of 0.25:

$$P^{0.25} \sim R^{0.5} \cdot T$$

Dividing by $R^{0.5}$ we can see that the temperature depends on the power output and radius in this way:

$$T \sim P^{0.25} / R^{0.5}$$

Since we already know that the power P grows with M^3 and the radius R with $M^{0.6}$, we can conclude the following:

$$T \sim M^{0.75} / M^{0.3}$$

$$T \sim M^{0.45}$$

So the surface temperature for main sequence stars grows with the mass to the power of 0.45.

$$\mathbf{(T' / T) = (M' / M)^{0.45}}$$

Fantastic, isn't it? I'm sure you didn't expect that stellar physics could be so simple and straight-forward. But let's not stop here. There's one more result I'd like to derive. The gravitational acceleration g (in m/s^2) at the surface of a body can be computed via this formula:

$$g = G \cdot M / R^2$$

with the gravitational constant G. You already saw this formula in the section "Escape Velocity". Since we know that the radius R of a star scales with $M^{0.6}$, we can write the following proportionality:

$$g \sim M / M^{1.2}$$

$$g \sim 1 / M^{0.2}$$

So for main sequence stars the gravitational acceleration experienced at the surface decreases with the star's mass.

$$\mathbf{g' / g = (M / M')^{0.2}}$$

However, the dependence is relatively weak. For the gravitational acceleration to halve, the mass of the star must increase by a factor of 30. Still, a surprising result. And just so you have a reference value: the gravitational acceleration at the surface of the sun is about $g = 274\ m/s^2$.

Now we are ready to go back to Sirius.

Let's find out how Sirius relates to our sun in terms of the radius, surface temperature and the gravitational acceleration. Again we have $M' = 2 \cdot M$.

For the radius we get:

$R' / R = 2^{0.6} \approx 1.5$

So the radius of Sirius is about 50 % larger than that of the sun. Let's turn to the surface temperature:

$T' / T = 2^{0.45} \approx 1.37$

The surface temperature is roughly 37 % higher.

$g' / g = (1/2)^{0.2} = 0.87$

The gravitational acceleration on Sirius is about 13 % lower.

If you want a challenge, use the formulas of this chapter to derive a relationship for the average density:

$D = M / V$

and escape velocity:

$v = \text{sqrt}(2 \cdot G \cdot M / R)$

of main sequence stars. If you do everything right, you should arrive at the result that the density is inversely proportional to $M^{0.8}$ and the escape velocity proportional to $M^{0.2}$ (remember that a square root means taking a quantity to the power of 0.5). And there's more. You could also take a look at the light intensity at the surface of the star:

$I = P / (4 \cdot \pi \cdot R^2)$

and central pressure:

$p \sim M^2 / R^4$

Try it out. Your results should be $I \sim M^{1.8}$ and $p \sim 1/M^{0.4}$. As you can see, main sequence stars provide a nice playground for anyone interested in mathematics and physics. The proportionality concept doesn't require a lot of prior knowledge and the results are intriguing.

Before going to the next section, let's take a quick look at the supergiants. In the luminosity - temperature coordinate system you can see that the line of supergiants is more or less parallel to the x-axis, meaning that their power output doesn't vary with temperature in first approximation. What can we deduce from that? Well, think Stefan-Boltzmann law: $P \sim R^2 \cdot T^4$. For the power not to vary with temperature, we need the terms on the right side to cancel each other:

$R^2 \sim 1 / T^4$

$R \sim 1 / T^2$

So hotter supergiants must be smaller. With this said, let's go back to earth. For great formulas, you don't need to look to the stars. Reaching for them though is never a bad idea.

- **Electrical Resistance**

Can you imagine a life without electricity? I bet most of us can't. We are used to having uninterrupted access to electricity for lighting, cooking, heating, telephones, computers, cars, ... It provides us with fantastic luxuries people one hundred years ago could only dream of. So it can't hurt to know the very basics.

Two quantities are central to any electric circuit: the voltage V (in V = volts), also called electric potential, and the electric current I (in A = amperes). You can take the word "current" literally as there is indeed something flowing in an electric circuit. But of course you know that there's no liquid or gas involved. It is the countless free electrons that make up this flow, driven by the electric potential that is provided by a battery for example. In very crude words, the voltage is the cause of the flow and the current its strength.

From this point of view, Ohm's law makes a lot of sense. It simple states that a strong flow (high electric current) implies a strong driving force (high voltage). The constant of proportionality in this relation is called electrical resistance R (in Ω = ohms):

$$\mathbf{V = R \cdot I}$$

So the electrical resistance is just the ratio of voltage to current. If you need to input a high voltage to produce just a little current, it means that the electric resistance in the circuit is high. What determines the value of this resistance? Well, suppose for a moment we were talking about water flow through a pipe. What are the factors that would hinder this flow?

One factor is the length of the pipe. The longer it is, the more friction we need to overcome to get the water from A to B. The same is true for electric flow. The longer a wire is, the higher its electric resistance. We denote the length by l (in m).

What else? Obviously the cross sectional area should be a factor. The bigger a pipe, the smaller the part of the water that is in contact with the boundaries (which cause friction) and the more water can be transported per second. Conductors of electricity work the same way. If a wire has a larger cross sectional area A (in m^2), its resistance will be smaller.

We can input these two quantities into this formula to determine the electric resistance of a conductor:

$R = r \cdot l / A$

The factor r (in Ωm) is called specific electrical resistance and depends on two things: the material of the conductor and its temperature T (in °C). Some conduct better when hot, others worse. We can estimate the value of the specific electrical resistance using this formula:

$r = r(0) \cdot (1 + \alpha \cdot (T - 20))$

with r(0) being the specific resistance at 20 °C and α the temperature coefficient. Both these quantities depend on the material of the conductor and can be found in tables. Now that's a lot of formulas, but they cover an enormous amount of physical problems as you will see in the examples.

A great and often used conductor is copper. Its specific electrical resistance at 20 °C and temperature coefficient is:

$r(0) = 1.7 \cdot 10^{-8}\ \Omega m$

$\alpha = 0.0039\ 1/K$

Note the positive value of α means that the resistance gets higher as the temperature increases. So in terms of conducting electricity, copper performs best at low temperatures. Compute the specific resistance at T = 35 °C and the overall resistance of a copper wire with length l = 6 m and cross section A = 0.00001 m² at this temperature.

First let's look at the specific resistance at 35 °C:

$r = 1.7 \cdot 10^{-8} \cdot (1 + 0.0039 \cdot (35 - 20))$

$r \approx 1.7 \cdot 10^{-8} \cdot 1.059$

(Quick remark before the result: the factor on the right side shows that the increase in resistance due to the elevated temperature is 5.9 %)

$r \approx 1.8 \cdot 10^{-8}\ \Omega m$

Now that we corrected the specific resistance for temperature, we can calculate the overall resistance of the copper wire:

$R = 1.8 \cdot 10^{-8}\ \Omega m \cdot 6\ m / 0.00001\ m^2$

$R \approx 0.011\ \Omega$

We want to make a thermometer for our oven. Using a volt and ampere meter, we manage to have the device display the current specific resistance of a copper wire in the circuit. Suppose the display outputs a specific resistance of $r = 2.9 \cdot 10^{-8}\ \Omega m$. *What is the current temperature in the oven? Use the reference values for copper from the first example.*

With the temperature formula we can set up an equation for the current temperature in the oven:

$$2.9 \cdot 10^{-8} = 1.7 \cdot 10^{-8} \cdot (1 + 0.0039 \cdot (T - 20))$$

First divide by $1.7 \cdot 10^{-8}$:

$$1.71 \approx 1 + 0.0039 \cdot (T - 20)$$

Now both sides minus one:

$$0.71 \approx 0.0039 \cdot (T - 20)$$

Divide by 0.0039 and then add 20 to both sides. This leads to the solution of the equation:

$$T \approx 202\ °C$$

As you can see, we can make smart use of the annoying dependence of resistance on temperature. We just inverse the formula to turn it into a way to measure temperature.

We want to use an aluminium wire (set $r = 2.8 \cdot 10^{-8}\ \Omega m$*) of length* $l = 5\ m$ *to produce a current of* $I = 900\ A$ *when applying* $V = 4.5\ V$. *What cross sectional area and radius should the wire have to accomplish this?*

From Ohm's law we can deduce that the overall resistance of the wire should be:

$R = V / I = 0.005\ \Omega$

With this value we can set up an equation for the cross sectional area A.

$R = r \cdot l / A$

$0.005\ \Omega = 2.8 \cdot 10^{-8}\ \Omega m \cdot 5\ m / A$

$0.005\ \Omega = 14 \cdot 10^{-8}\ \Omega m^2 / A$

Multiply both sides by A:

$0.005\ \Omega \cdot A = 14 \cdot 10^{-8}\ \Omega m^2$

Divide by 0.005 Ω:

$A = 0.000028\ m^2$

What radius will produce this cross section? With the area formula for circles $A = \pi \cdot R^2$ *we get:*

$0.000028\ m^2 = \pi \cdot R^2$

Divide by π and apply the square root:

$R \approx 0.003\ m = 3\ mm$

This was just a quick peek into the rich and unending field of electricity. The next step you could take is to learn how to add electrical resistance in a circuit properly. The "Tipler" (very pricey, but also very thorough) is a good place to start.

- **Strings and Sound**

I suppose at one point we've all found ourselves stretching and plucking a string to produce sounds of different frequencies. The more force you applied in stretching the string, the higher the pitch of the resulting sound. Strings on musical instruments basically work the same way and calculating the pitch is actually quite simple. The great formula in this section will allow us to do just that.

What variables do we need? Obviously we need the force used to stretch the string, that is, the tension T (in N). Additionally, we need the length of the string L (in m) and the linear density D (in kilogram per meter string, kg/m). Note that with string length we only mean the vibrating part of the string, the rest we don't care about. The resulting fundamental frequency of the string f (in Hz) can be calculated by this formula:

f = (0.5 / L) · sqrt (T / D)

Let's take a look at two examples.

A steel string has a linear density of about D = 0.06 kg/m. Assume we stretch it with a force of T = 3000 N to a length of L = 0.5 m. What is the resulting fundamental frequency?

f = (0.5 / 0.5 m) · sqrt (3000 N / 0.06 kg/m)

f ≈ 224 Hz

This corresponds roughly to the note A one octave below the chamber pitch.

Assume we want to tune said string to the chamber pitch $f = 440$ *Hz. What force do we need to apply? Using the given values we can set up an equation (I'll leave the units out):*

$440 = (0.5 / 0.5) \cdot sqrt (T / 0.06)$

$440 = sqrt (T / 0.06)$

Squaring both sides leads to:

$193{,}600 = T / 0.06$

Now multiply both sides by 0.06:

$T \approx 11{,}600\ N = 11.6\ kN$

Is this much? Well, it's the same as the gravitational force of an average car, so yes, it indeed is a lot. Now you know why the strings needs to be made of a tough material such as steel.

How does the frequency vary as we change the parameters? Let's take a look at the inputs separately:

- If we halve the length of the string, the frequency will double (thus producing a higher pitch). That's why playing musical instruments such as the guitar, bass violin, ... works. By pushing the finger down on the string, the musicians alters the length of the vibrating part and thus the pitch.

- If we quadruple the tension, the frequency doubles. To produce a sound one octave above the chamber pitch (f = 880 Hz) with the above string, we would need to apply a force of roughly T = 4 · 11.6 kN = 46.4 kN.

- If we quadruple the linear density, for example by doubling its diameter (more mass per meter), the frequency halves. So thicker strings produce a lower pitch. This is probably not news to anyone playing a stringed instrument.

As you can see, even without plugging in values, the formula tells a vivid story. We can deduce a lot by simply analyzing its structure. One last remark: why say fundamental frequency and not just frequency? I'm glad you asked.

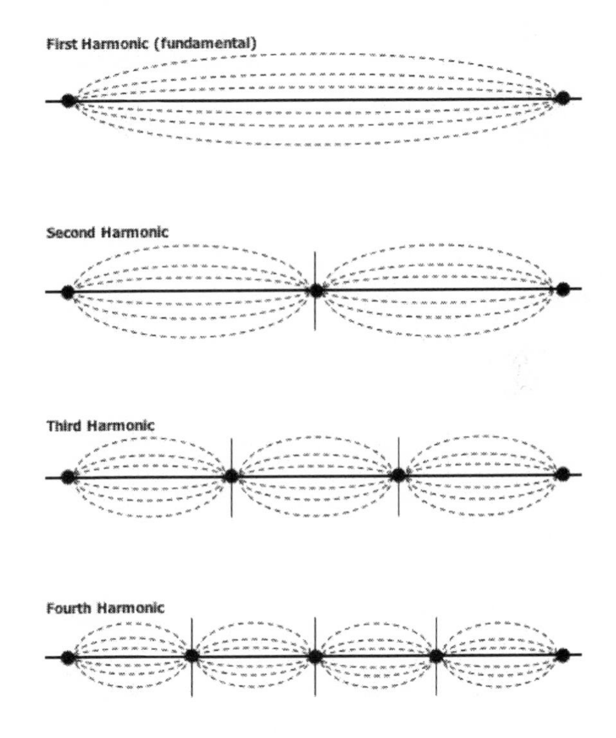

A string never produces one single frequency. For any length and tension, we can have many different vibrating modes, all producing different frequencies. The picture above shows four basic vibrating modes.

In the most basic vibrating mode the string oscillates up and down as a whole. This is the mode that produces the fundamental frequency f (also called the first harmonic). In all other modes, the so called nodes appear. These are stationary points on the string at which the string stands still.

The second harmonic is produced with one node in the middle. Since it effectively halves the string, the frequency doubles to $2 \cdot f$ (one octave higher). The appearance of a second node cuts the string in three parts and accordingly it vibrates at a frequency $3 \cdot f$ (another octave higher). This is the third harmonic. Further harmonics are added by the same principle.

It's noteworthy that the amplitude, which for our purpose is just the volume, of the produced sound decreases with each added node. This is why these overtones are not consciously noticeable. Despite that, they play a very important role in music. It needs the harmonics to sound rich and natural. Cut them out, for example by using a low-pass-filter, and the music will sound bland and artificial.

Part II: Mathematics

- **Cylinders**

A very fundamental and important geometric shape is the cylinder. There are plenty of real-world examples of cylindrical objects: tree trunks, towers, water tanks, barrels, rods, pistons, wires, antennas, telescopes, cans, buckets, coins, drums, candles, and so on. To characterize a cylinder you need two quantities: the radius r (in m) and the height h (in m). Of course you can also use different units, in which case you should make sure that you use the same unit for both quantities.

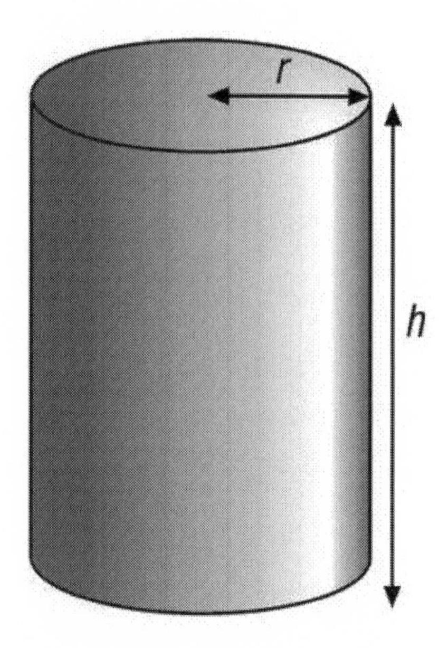

Let's talk about the volume V (in m^3) first. We can calculate it by multiplying the base area with the height. Since the base is just a circle with the area $\pi \cdot r^2$, we can write:

$$\mathbf{V = \pi \cdot r^2 \cdot h}$$

That was painless. What about the surface area? For the moment, let's ignore the circular base and top areas and focus on the mantle. Imagine this mantle rolled out. It would then turn into a rectangle with the height as one side and the circumference of the circle, which is $2 \cdot \pi \cdot r$, as as the other side. Since the area for a rectangle is just the product of the two sides, we get this expression for the area of the mantle surface:

$$\mathbf{M = 2 \cdot \pi \cdot r \cdot h}$$

If we add the areas of the mantle, base and top, we get the total surface area of a cylinder:

$$S = 2 \cdot \pi \cdot r \cdot h + 2 \cdot \pi \cdot r^2$$

$$\mathbf{S = 2 \cdot \pi \cdot r \cdot (h + r)}$$

So despite the curved shape, we were able to easily derive a formula for the mantle and surface area. All that was necessary was to imagine the mantle being rolled out and stubbornly draw the conclusions.

A useful addition here is the formula for calculating the mass. In many geometry problems you are given a density D (in kg/m^3) of the material the cylinders is made of and asked to compute the mass m (in kg). For that you can use this formula:

$$\mathbf{m = D \cdot V}$$

So the mass is simply the product of density and volume. Sometimes, for example in the case of thin metal sheets, it makes more sense to give a "two-dimensional density", that is, the density in kilograms per square meter rather than per cubic meter. We denote this density by D' (in kg/m^2). In this case we calculate the mass by this formula:

$$m = D' \cdot A$$

with the area A either being only the mantle or the entire surface area. This formula is useful when dealing with hollow cylinders like cans, where the only source of mass is the sheet that makes up the surface (at least as long as it's empty).

Before we move on to an interesting optimization problem involving cylinders, let's do some examples.

What is the volume, mantle area and surface area of a pine tree log with radius $r = 0.6$ m and $h = 5$ m? What is its mass? The density of pine wood is $D = 500\ kg/m^3$.

First let's calculate the volume:

$$V = \pi \cdot (0.6\ m)^2 \cdot 5\ m \approx 5.65\ m^3$$

The mantle and surface area is:

$$M = 2 \cdot \pi \cdot 0.6\ m \cdot 5m \approx 18.85\ m^2$$

$$S = 18.85 + 2 \cdot \pi \cdot (0.6\ m)^2 \approx 21.11\ m^2$$

Now onto the mass. Remember that to compute the mass, we simply multiply the density by the volume:

$$m = 500\ kg/m^3 \cdot 5.65\ m^3 \approx 2825\ kg$$

That's about the weight of three small cars.

We make a can out of a thin aluminium sheet with the density $D = 2.7 kg/m^2$ (which corresponds to 1 mm thickness). What is the mass of such a can assuming $r = 0.03 m$ and $h = 0.15 m$ and ignoring the opening at the top?

Note that we are given a two dimensional density here. So the the mass must be determined from the surface area. Calculating it is the first step:

$S = 2 \cdot \pi \cdot 0.03 m \cdot 0.18 m \approx 0.034 m^2$

So the mass of the (empty can) is:

$m = 2.7 kg/m^2 \cdot 0.034 m^2 \approx 0.092 kg = 92 g$

We know that the volume of a cylinder is $V = 10 ft^3$ and its mantle surface $M = 14 ft^2$. What are the dimensions of the cylinder? This is not an easy one. From the given values we can set up two equations:

(a) $\pi \cdot r^2 \cdot h = 10$

(b) $2 \cdot \pi \cdot r \cdot h = 14$

There are several ways to approach this. The (in my opinion) easiest way is to do the ratio of the two equations, so we divide (a) by (b). Because the height appears as a linear factor in both equations, it will cancel out, leaving an equation for the radius only. So we divide:

$\pi \cdot r^2 \cdot h / (2 \cdot \pi \cdot r \cdot h) = 10 / 14$

Now cancel out h and π:

$r^2 / (2 \cdot r) = 10 / 14$

We can further simplify that:

$r / 2 = 10 / 14$

And multiply by 2:

$r = 20 / 14 \approx 1.43 \, ft$

Since all the inputs are in feet, we get the radius in feet as well. Now that we know the radius, we can use any of the two equations (a) and (b) to determine the height.

$\pi \cdot (1.43)^2 \cdot h = 10$

$6.42 \cdot h \approx 10$

Divide by 6.42:

$h \approx 1.56 \, ft$

Done! So the dimensions of the cylinder we were looking for are r = 1.43 ft and h = 1.58 ft. Let's see if it all checks out:

$V = \pi \cdot (1.43 \, ft)^2 \cdot 1.56 \, ft \approx 10 \, ft^3$

$M = 2 \cdot \pi \cdot 1.43 \, ft \cdot 1.56 \, ft \approx 14 \, ft^2$

It all checks out very nicely.

There's a great optimization problem that goes along with cylinders. Suppose you manufacture cans with a given volume. What dimensions should you choose? Well, obviously it would be great to have the surface area as small as possible. This way you can keep the material costs at a minimum. So given the volume V the can should have, what dimensions produce the minimum surface area?

Since the solution of this problem requires some calculus, I will only give the first step here. Express the cylinder's surface area as a function of the volume and radius:

$h = V / (\pi \cdot r^2)$

$S(r) = 2 \cdot \pi \cdot r \cdot (V / (\pi \cdot r^2) + r)$

After some simplification:

$S(r) = 2 \cdot (V / r + \pi \cdot r^2)$

Now you can find the minimum of the function by doing the first derivative and finding the root. This leads to the optimal radius and optimal height:

$\mathbf{r(opt) = (0.5 \cdot V / \pi)^{0.33}}$

$\mathbf{h(opt) = 2 \cdot r(opt)}$

Note that two times the radius is just the diameter of the cylinder. So a cylinder with the smallest possible surface area is always equal in height and diameter (think soup cans). Let's make an optimal can.

We want to manufacture an optimized can that holds $V = 200\ ml = 200\ cm^3$. These are the dimensions we should choose:

$r(opt) = (0.5 \cdot 200 / \pi)^{0.33} \approx 3.13\ cm$

$h(opt) \approx 2 \cdot 3.13$ cm = 6.26 cm

With these dimensions the surface area is:

$S = 2 \cdot \pi \cdot 3.13$ cm $\cdot$ 9.39 cm $\approx$ 184.67 cm^2

For the given volume, it can't go below that.

Hopefully this quick overview of cylinders was helpful and informative. In the next section we are going to continue with yet another fundamental and important shape: the triangle.

- **Arbitrary Triangles**

If you read the first volume carefully, you probably master right triangles by now. But why always limit yourself to having a right angle? After all, many triangles occurring in nature don't offer the luxury of a right angle. Aren't there any formulas for triangles in general? Luckily there are. And this section is all about those.

Before we can state the formulas, let's talk conventions. The corners of a triangle are named A, B and C in an anti-clockwise fashion. The angle spawned at the corners are denoted by α, β and γ respectively. Finally, the side opposite the corner is named a, b and c respectively. Take a look at the image below to make sure you understood these conventions. It'll help to apply the formulas.

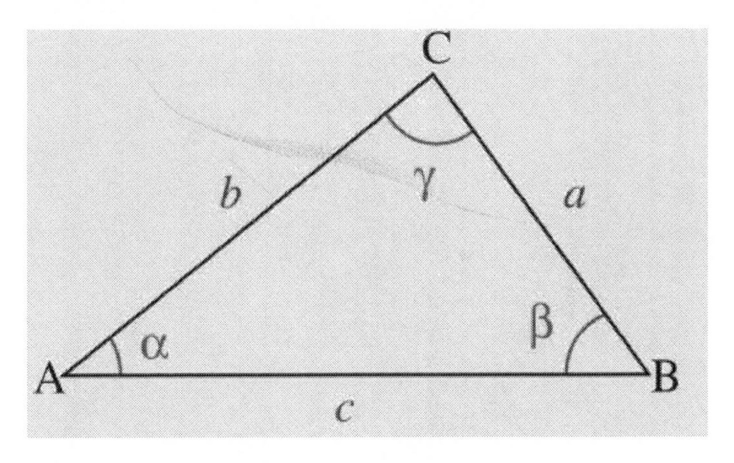

Now onto the first law, called the law of sines. It is the formula used for a technique called triangulation. The law states that the ratio of a side to the sine of the angle opposite to it is a constant. In mathematical terms:

$a / \sin(\alpha) = b / \sin(\beta) = c / \sin(\gamma)$

For example when given two angles (let's say α and β) and one side (for example a), you can use it to determine the length of a remaining side (in this case b). Let's put the formula to use.

Two ships, A and B, spot a third ship C at the respective angles $\alpha = 20°$ and $\beta = 50°$. From GPS measurements we know that B and C are $a = 12$ miles apart. What's the distance between ships A and C?

Look at the diagram (or better yet, make your own diagram) to see what is given and what is needed. Obviously we want to compute side b. We can set up an equation for it using the law of sines:

$$12 / \sin(20°) = b / \sin(50°)$$

$$35.09 = b / \sin(50°)$$

Make sure your calculator is set to degrees and not radians to get the correct result. Multiply both sides by $\sin(50°)$:

$$b = 35.09 \cdot \sin(50°) \approx 26.88 \text{ miles}$$

This example comes from the website analyzemath.com. You are standing at an unknown distance from a building and observe it at an angle of 50°. Now you walk 30 m towards it and observe it an angle of 60°. What's the height h of the building? The image below is a visualization of this situation.

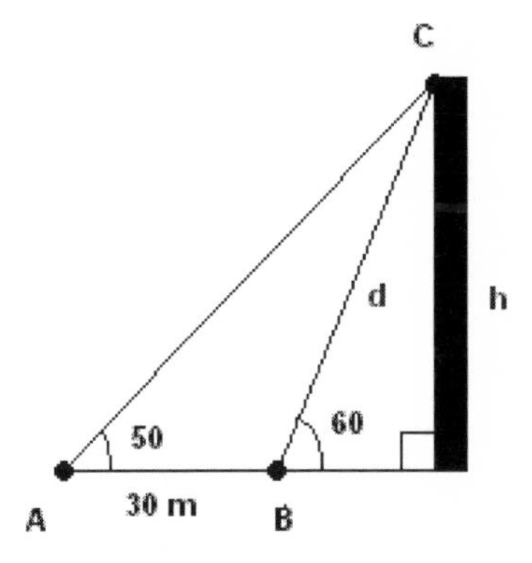

The law of sines does not allow us to directly compute h. But it does allow us to determine the length of d, which is the line of sight to the top of the building after walking the 30 m. This in turn will enable us to get the height.

To apply the law of sines, we need the angle opposite to side c, which is the 30 m. Can we deduce that? Yes, we can. The angles in any triangle must add up to 180 °. The angle at point A is 50°, the angle at point B 120°. Thus at point C, the angle must be 10° for everything to check out. Now let's set up the equation for d:

$30 / \sin(10°) = d / \sin(50°)$

$172.76 = d / \sin(50°)$

$d = 172.76 \cdot \sin(50°) \approx 132.34 \text{ m}$

Now let's focus on the right triangle that includes the height h. This side is the opposite to the known angle (60°), d is the hypotenuse of the triangle. We can use the common sine formula to compute the height:

$\sin(50°) = h / 132.34$

$h = 132.34 \cdot sin(50°) \approx 101.38\ m$

That was tough! But without the law of sines, determining the height would not have been possible (well, ok, you could have just asked at the reception).

There's another law that can be used when dealing with arbitrary triangles. It is basically the generalization of Pythagoras' theorem. You can use it whenever you know two sides of the triangle and the angle between them. Assuming you know the sides a, b and the angle γ between them, you can use the law of cosines to calculate the remaining side c:

$$\mathbf{c^2 = a^2 + b^2 - 2 \cdot a \cdot b \cdot cos(\gamma)}$$

For a right angle between the given sides ($\gamma = 90°$), this reduces to the well-known Pythagoras formula $c^2 = a^2 + b^2$. So the law of cosines contains this special case.

A radar station located at point C tracks two planes currently located at A and B. Seen from the radar station, the angle between them is $\gamma = 75°$. Their respective distance to the radar station is a = 14 miles and b = 6 miles. Draw a sketch to visualize this situation. How far apart are the planes?

We are given side a and b as well as the angle between them. The distance between the planes corresponds to side c. To calculate it, we can directly apply the law of cosines:

$c^2 = 14^2 + 6^2 - 2 \cdot 14 \cdot 6 \cdot cos(75°)$

$c^2 \approx 188.52$

c ≈ sqrt (188.52) ≈ 13.73 miles

Using the law of sines we could now determine the remaining angles. However, in this application we're not interested in doing that.

From a triangle we only know that it has the side lengths a = 10, b = 20 and c = 25. Can we determine all the angles in the triangle from that? Indeed we can and will. Since the law of sines only works when we know at least one angle, we must start with the generalized form of Pythagoras' theorem. This will lead to the angle γ, which is opposite side c.

$25^2 = 10^2 + 20^2 - 2 \cdot 10 \cdot 20 \cdot \cos(\gamma)$

$625 = 500 - 400 \cdot \cos(\gamma)$

Subtract 500 from both sides:

$125 = -400 \cdot \cos(\gamma)$

Divide by -400:

$-0.31 \approx \cos(\gamma)$

Using the inverse cosine function we find:

$\gamma \approx \arccos(-0.31) \approx 108.1°$

So now we already know one of the angles. We could continue with the law of cosines, but using the law of sines is easier from here on. Let's determine β, which is opposite side b.

$b / \sin(\beta) = c / \sin(\gamma)$

$20 / sin(\beta) = 25 / sin(108.1°)$

$20 / sin(\beta) = 26.30$

Multiply by $sin(\beta)$:

$20 = 26.30 \cdot sin(\beta)$

Divide by 26.30:

$0.76 = sin(\beta)$

Again we use our calculator's inverse function to find the value of the angle. This leads to:

$\beta = arcsin(0.76) \approx 49.5°$

So now we already know two of the three angles. We could apply the law of sines one more time to find the remaining angle. But there's a more elegant way to get it. All angles in the triangle must add up to 180°, so the remaining angle must be:

$\alpha = 180° - 108.1° - 49.5° = 22.4°$

It was a lot of work, but we were able to determine all the angles from the lengths of the sides.

Arbitrary triangles are not so easy to handle as right triangles, but with the law of sines and the law of cosines it is possible to tame them. If you want to master geometry, you should try your best to become familiar with them. And if you get bored with those, the next step would be to look at geometry on spherical surfaces.

- **Summation**

Carl Friedrich Gauß was without a doubt one of the greatest mathematicians that ever lived. His genius became apparent at a very young age. One day, so the tale goes, his primary school teacher asked the class to sum all numbers from 1 to 100. Why he did that I can't say, maybe he just wanted to keep the kids occupied, but he most certainly was very surprised when young Carl gave the correct answer after just a few seconds of thought: 5050. I guess it takes more to keep a genius occupied.

How did Gauß arrive at the answer so quickly? Instead of starting with $1 + 2 + 3$ and continuing this way, which is what most of us would do, he noticed that the sum is much easier to do when pairing the highest and lowest addends:

$1 + 100 = 101$

$2 + 99 = 101$

$3 + 98 = 101$

If we keep doing this, at some point we will arrive at the last pair, which is:

$50 + 51 = 101$

So the sum of all numbers from 1 to 100 comes down to a single multiplication: $50 \cdot 101 = 5050$. Young Carl saw that immediately and went on to be a great mathematician. His accomplishments earned him a spot on the 10 Deutsche Mark bill, including references to his achievements in statistics, land surveying and astronomy (see image below).

We can use his insight to derive a general formula for this case. Suppose we want to add all numbers from 1 to a certain number n. For now assume n to be even, but luckily the resulting formula will hold true for odd numbers as well. Let's order the addends as Gauß did:

$1 + n = n + 1$

$2 + (n-1) = n + 1$

$3 + (n-2) = n + 1$

...

$0.5 \cdot n + (0.5 \cdot n + 1) = n + 1$

As before, the summation comes down to a single multiplication. This provides us with a truly great and on top of that surprisingly useful formula:

$$\mathbf{1 + 2 + ... + n = 0.5 \cdot n \cdot (n + 1)}$$

Let's turn to the examples.

What is the sum of all numbers between 101 and 200? So we want to know the value of this sum:

$$S = 101 + 102 + ... + 200$$

Since the sum doesn't start with 1, we can't directly apply the formula. However, let's look at the big picture. We should arrive at the correct result when we do the sum of all numbers from 1 to 200 in a first step, and subtract from that the sum of all numbers from 1 to 100 in a second step.

$$1 + 2 + ... + 200 = 0.5 \cdot 200 \cdot 201 = 20{,}100$$

$$1 + 2 + ... + 100 = 0.5 \cdot 100 \cdot 101 = 5050$$

This leads to:

$$S = 20{,}100 - 5050 = 15{,}050$$

So to put the addition formula to use, it is not necessary for the sum to start at 1. If we apply it in two steps, the sum can start and end anywhere.

The great summation formula is vital in solving the birthday problem (mathematicians seem to be grumpy people, they have a problem with everything, even birthdays). The question is: how likely is it that at least two people in a group of n people share a birthday?

It's easier to look at the opposite event first, that is, the chance that no two people share a birthday. In my book "Statistical Snacks" I analyzed this problem and arrived at this result:

$p(\text{no shared birthdays}) = (364/365)^m$

with m being the possible number of pairs in the group of n. But this doesn't seem to be a very helpful result, we still need to find a relationship between group size and the number of pairs. Surprisingly, this is where Gauß comes in.

Let's assign all the members of the group a number: person 1, person 2, and so on up to person n. Person 1 can enter into n-1 pairings. Person 2 can also enter into n-1 pairings, but since we already covered his pairing with person 1, he only adds n-2 pairings to the total. Person 3 also has a number of n-1 pairings available, but since we covered his pairing with person 1 and person 2, he only adds n-3 pairings to the total. In this way we can continue to add up the total number of pairs, until we get to person n, who won't contribute anything to the total.

$m = n-1 + n-2 + n-3 + ... + 0$

We can write the 0 as n-n. This leads to:

$m = n-1 + n-2 + n-3 + ... + n-n$

Let's rearrange this a bit (numbers with a plus-sign to the left, numbers with a minus-sign to the right):

$m = n + n + ... + n - 1 - 2 - 3 - ... n$

$m = n \cdot n - (1 + 2 + 3 + ... + n)$

In the bracket the sum of all numbers from 1 to n appears. Thus, we can insert the summation formula:

$m = n \cdot n - 0.5 \cdot n \cdot (n + 1)$

Now we can easily calculate the number of pairs m from the group size n and the previous result for the probability of having no shared birthdays becomes useful. So, how likely is it to have no shared birthdays in a group of 50 people?

In a group of n = 50 people we have this number of pairs:

$m = 50 \cdot 50 - 0.5 \cdot 50 \cdot 51 = 1225$

The chance of having no shared birthdays is thus:

$p(\text{no shared birthdays}) = (364/365)^{1225}$

$p(\text{no shared birthdays}) \approx 0.03 = 3\ \%$

In other words: there's a solid 97 % chance that two people in this group share a birthday, practically a sure thing so to say. Surprised? Most people are when the first get some numbers on the birthday problem.

If you didn't believe it before, I'm sure you'll now agree with me that pure mathematics can be quite interesting. Aside from being intellectually stimulating, the results can be applied to problems that seem to have no relation to the initial train of thought. I'm sure Gauß did not think about the birthday problem when he found a way to sum more efficiently, but it was this that allowed the problem to be solved in the end.

- **Standard Deviation and Error**

This is an excerpt from "Statistical Snacks" by Metin Bektas.

Let's take a look at an important statistical quantity. The standard deviation tells us how strongly observed values deviate from their averaged value. It also allows us to say how likely it is that further measurements will fall into a certain interval around the mean. Let's go through an example of calculating the standard deviation and drawing conclusions from that.

When you buy a food product, you can find the weight of the contents on the box. One manufacturer for example claims that his box of cookies weighs 100 grams. To test the reliability of this claim, we take a sample of 10 boxes and weigh them. Here are the results: 102 gr, 105 gr, 98 gr, 101 gr, 94 gr, 98 gr, 103 gr, 105 gr, 101 gr and 97 gr.

Our first job will be to calculate the arithmetic mean, usually symbolized by μ. To do that, we simply add all the values and divide by the number of measurements:

$\mu = 1004 / 10 = 100.4$ grams

So the mean indeed was very close to the 100 grams claimed by the manufacturer. But additionally to that, we are interested in knowing how strongly the weights spread around this mean. How likely is getting a 90 gram box for example? Is that something we can expect to happen often?

To get a sense of the spread we turn to the standard deviation. Let's symbolize the measured values with m. For each of these values we compute this quantity:

$x = (m - \mu)^2$

It is simply the square of the difference between the observed and mean value. Once we did that for all the measurements, we can get the standard deviation from this formula:

$s = \text{square root} ((x_1 + x_2 + \ldots) / (n - 1))$

with n being the total number of measured values (in our case 10). Let's calculate the standard deviation now for the weights we observed. For the first and second measurement we get:

$x_1 = (102 - 100.4)^2 = 2.6$

$x_2 = (105 - 100.4)^2 = 21.2$

We proceed in a similar fashion to get the corresponding results for the other 8 measurements. Once this is done, we sum them all up. The sum turns out to be 116.6. Plugging that value into the above formula for the standard deviation results in:

$s = \text{square root} (116.6 / 9) = 3.6$ grams

This is the standard deviation of our sample. If the quantity of interest, which for us is the weight, is normally distributed (an assumption we can make if we don't have any further information), one can use this table to draw conclusions:

- 68.3 % chance of being within $\mu \pm s$

- 95.5 % chance of being within $\mu \pm 2 \cdot s$

- 99.7 % chance of being within $\mu \pm 3 \cdot s$

In the example we have:

$\mu + s = 104.0$ grams

$\mu - s = 96.8$ grams

If our sample is representative, we can expect 68.3 % of all further measurements to be within this interval. This is what's called the “1 sigma interval”. It covers one standard deviation around the mean.

For the “2 sigma interval” we take two standard deviations around the mean. Using the calculated values we get:

$\mu + 2 \cdot s = 107.6$ grams

$\mu - 2 \cdot s = 93.2$ grams

Our estimate thus is that 95.5 % of all boxes fall within this range. As you can see, the standard deviation indeed does provide a lot of information about how the weights spread.

As mentioned, these conclusions are only reliable if our sample is representative, which means that it should be large and have no systematic errors. In our example we shouldn't be too confident as 10 measurements is not a significant sample. And if we used a cheap scale to weigh the boxes or handled the scale incorrectly (for example not properly setting it to zero before starting the experiment), a systematic error is a possibility. So make sure to consider that when looking at the mean and standard deviation of a sample.

I also mentioned that the above table only holds exactly true when the quantity in question is normally distributed. If the quantity is distributed in another way, we can only use the table as a first approximation. As a rule of thumb one can say that when the observed values are strongly skewed (significantly more than half of the observed values left or right of the mean), one should not assume a normal distribution.

The standard deviation should not be confused with the standard error SE, which rather measures how reliable our result for the mean is. Remember that we got $\mu = 100.4$ grams as the mean for our sample. Since our sample was small, there's a good chance that the true mean μ(true) will differ somewhat from that. Can we say to what extend?

To do that, we calculate standard error by dividing the standard deviation by the square root of the number of measurements:

SE = s / square root (n)

Plugging in the corresponding values for the boxes of cookies results in SE = 1.14. Now what does that tell us? We can conclude (and this is a general rule) that the true mean is within twice the standard error with a chance of 95 %. For us this means that there's 95 % certainty that the true mean of the box weights lies within this interval:

μ(true) = 100.4 ± 2.28 gr

Despite only having a sample size of 10 boxes, we were able to narrow the mean down to a relatively small interval. Note how the standard error varies with the sample size n. Assuming the standard deviation stays relatively constant when expanding the sample, the standard error will halve when the sample size increases fourfold. With 40 boxes we could bring the interval down to about ± 1.14 grams.

So remember that while the standard deviation gives information about the spread around the mean, the standard error helps us to deduce how close our sample mean is to the true mean. Both is helpful information to have for any sample

- **Zipf Distribution**

Common dictionaries list the words of a language in alphabetical order. However, there's a much more informative way to order them, that is, by frequency of usage. We put the most commonly used word on rank 1, the second most commonly used on rank 2, and so on. Of course compiling such a dictionary requires an enormous amount of statistical analysis and luckily computers have made this a lot easier. But the first frequency dictionary actually came long before the age of computers. In 1898 a comprehensive frequency dictionary was published for the German language, other languages followed soon.

Before we go on, let's define what we mean by the frequency of an event (in a non-physical sense). If an event occurs n times among N trials, its frequency is $f = n / N$. For example, the most common English word "the" appears on average 7 times among 100 words, so its frequency is $f = 0.07 = 7\%$.

Back to the frequency dictionary. Assume we write next to each word its frequency. Since the words are ordered by usage, the frequency will obviously decline as we go to higher ranks. George Kingsley Zipf, called the father quantitative linguistics by some, made an intriguing discovery. If you multiply the rank with the according frequency, you always end up with roughly the same number. In other words: the frequency f of usage is inversely proportional to the rank r.

$f \cdot r = \text{const.}$

$\mathbf{f = C / r}$

with a constant C. You might dismiss this as an oddity, but

this law, appropriately called Zipf's law, holds true in many cases when ordering elements by a certain quantity such as frequency, magnitude or size. For example, when you order cities by their size s (either measured in number of people or area), again you can find out that the size varies inversely proportional to rank r:

$s = C / r$

You get a very similar result when ordering islands or deserts by their size. Recently I discovered that Zipf's law also holds true for my WordPress blog "Metin's Media & Math". When you order all the blog posts by the number of views v, the product of views and rank r remains constant:

$v = C / r$

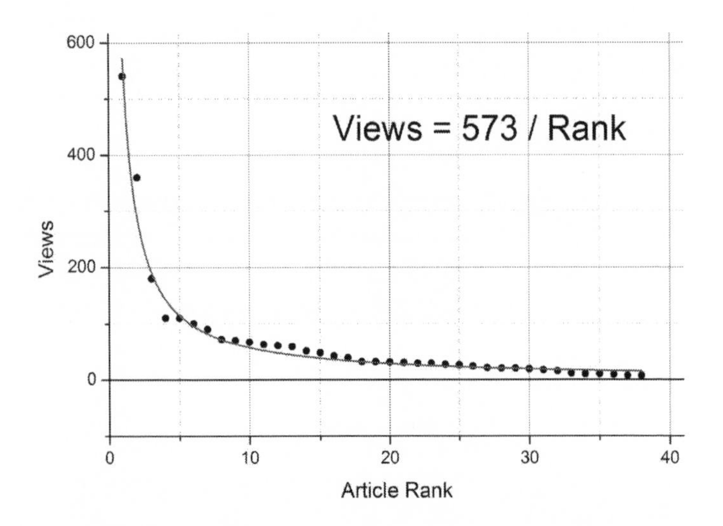

A similar relationship is found when looking at the number of views ordered by countries. A little research easily brings forth more and more examples of Zipf's law: ordering companies by staff, universities by number of students, world languages by number of speakers, solar flares by energy, oil fields by size, websites by hits, and so on. It seems to be a distribution that naturally occurs in all kinds of systems.

A helpful addition and a tool to draw neat conclusions from the Zipf distribution is the summation formula for the harmonic series, as presented in first volume of this book. Sometimes we need to do a sum over ranks. In this case the harmonic series appears and we can use this approximation formula to compute it (ln is the natural logarithm):

$$1/1 + 1/2 + 1/3 + ... + 1/r \approx \ln(r) + 0.58$$

For the English language the constant of proportionality in Zipf's law is about 0.075. Thus, we have this relationship between frequency and rank:

$$f = 0.075 / r$$

We will use this to answer two interesting questions. The first question is: what percentage of a typical English text is made up of the 1000 most commonly used words? To answer that, we have to add the frequencies from $r = 1$ to $r = 1000$:

$$f = 0.075 \cdot (1/1 + 1/2 + ... + 1/1000)$$

Summing all these terms by hand would be madness, this is where the approximation formula for the harmonic series comes in handy.

$$f \approx 0.075 \cdot (\ln(1000) + 0.58) \approx 0.56 = 56 \%$$

So the top 1000 words make up a little more than one half of a typical English text. Did you expect this result? Let's turn to another question: how many English words are there in total? Of course we can only hope to give a rough answer, but this won't stop us from doing the calculation.

Assume there are a total of N words in the language. So when we repeat the above calculation to find out what percentage of a common English text consists of these N words, the answer better be 100 %. Or in mathematical terms:

$$f = 0.075 \cdot (ln(N) + 0.58) = 1 = 100 \%$$

This is an equation for N. All that's left is some algebra. First we divide both sides by 0.075:

$$ln(N) + 0.58 \approx 13.3$$

Subtract 0.58:

$$ln(N) \approx 12.8$$

Now apply Euler's number, the inverse to the natural logarithm, to both sides. This leads to:

$$N \approx e^{12.8} \approx 350,000 \text{ words}$$

I rounded generously since this is not supposed to be more than a very rough estimate. How close is this to the real deal? That's hard to tell because exact numbers are not known. However, most estimates put the total number of words somewhere around 300,000 words, so the result seems realistic.

For e-books on Amazon the relationship between the daily sales rate s and the rank r is approximately given by:

s = 100,000 / r

So a book on rank r = 10,000 can be expected to sell s = 10 copies per day. As of November 2013, there are about 2.4 million e-books available on Amazon's US store (talk about a tough competition). Again we'll answer two questions. The first one is: how many e-books are sold on Amazon each day? To answer that, we need to add the daily sales rate from r = 1 to r = 2,400,000.

s = 100,000 · (1/1 + 1/2 + ... + 1/2,400,000)

Again we turn to the summation formula:

s ≈ 100,000 · (ln(2,400,000) + 0.58) ≈ 1.5 million

That's a lot of e-books! And a lot of saved trees for that matter. What percentage of the e-book sales come from the top 100 books? Have a guess before reading on. Let's calculate the total daily sales for the top 100 e-books:

s ≈ 100,000 · (ln(100) + 0.58) ≈ 0.5 million

Thus, the top 100 e-books already make up one-third of all sales while the other 2,399,900 e-books have to share the remaining two-thirds. The cake is very unevenly distributed.

In the final example I want to derive a formula which allows you to estimate the number of views for the most popular post on a blog. If you rank the posts of a blog by number of views v, the result is a Zipf distribution:

v = C / r

The most popular post is the one on rank one. Inserting $r = 1$, we can see that the number of views for this post (let's denote it by v') coincides with the value of the constant:

$$v' = C$$

Assume the blog has n entries in total. Then the total number of views V is given by this expression:

$$V = C \cdot (1/1 + 1/2 + ... + 1/n)$$

$$V \approx v' \cdot (\ln(n) + 0.58)$$

Thus, the number of views for the most popular entry can be estimated using this formula:

$$\mathbf{v' \approx V / (\ln(n) + 0.58)}$$

For example, if a blog accumulated $V = 10{,}000$ views and consists of $n = 120$ entries, the number one post will have this number of views if the views are distributed according to Zipf's law:

$$v' \approx 10{,}000 / (\ln(120) + 0.58) \approx 1900 \text{ views}$$

Of course you should enjoy all these results with a plus/minus 10 % accuracy since no ranking will lead to a perfect Zipf distribution. But for most purposes, this is good enough.

I hope this short bit helped you to appreciate Zipf's law. It is an all-rounder and appears almost every time we rank items. Combined with the summation formula for the harmonic series, we can use it to deduce intriguing results.

Part IV: Appendix

- **Unit Conversions**

Since we often need to convert units from the United States customary system (USCS) to the metric (SI) system and vice versa, here's a list of commonly needed conversion factors.

Lengths, SI to USCS:

Multiply meters with 3.28 to get to feet

- Multiply meters with 1.09 to get to yards
- Multiply meters with 0.00062 to get to miles
- Multiply kilometers with 3281 to get to feet
- Multiply kilometers with 1094 to get to yards
- Multiply kilometers with 0.62 to get to miles

Lengths, USCS to SI:

- Multiply feet with 0.30 to get to meters
- Multiply feet with 0.00030 to get to kilometers
- Multiply yards with 0.91 to get to meters
- Multiply yards with 0.00091 to get to kilometers
- Multiply miles with 1609 to get to meters
- Multiply miles with 1.61 to get to kilometers

To convert a squared to a squared unit, use the square of the conversion factor. For example you multiply m^2 by $3.3^2 \approx 10.9$ to get to ft^2. In a similar fashion, using the cube of the conversion factor, you can convert cubed units.

Speeds:

- Multiply m/s with 3.6 to get to km/h
- Multiply m/s with 2.23 to get to mph
- Multiply km/h with 0.28 to get to m/s
- Multiply km/h with 0.62 to get to mph
- Multiply mph with 0.45 to get to m/s
- Multiply mph with 1.61 to get to km/h

Other commonly used units:

- Multiply pounds with 0.45 to get to kilograms
- Multiply kilograms with 2.22 to get to pounds
- 1 liter = 0.001 m^3
- Multiply liters with 0.62 to get to gallons
- Multiply gallons with 3.79 to get to liters
- Celsius to Fahrenheit: °F = 1.8 · °C + 32
- Fahrenheit to Celsius: °C = 5/9 · (°F - 32)
- Kelvin to Celsius: °C = K - 273.15
- Celsius to Kelvin: K = °C + 273.15

- **Unit Prefixes**

peta (P) = 1,000,000,000,000,000 = 10^{15}

tera (T) = 1,000,000,000,000 = 10^{12}

giga (G) = 1,000,000,000 = 10^{9}

mega (M) = 1,000,000 = 10^{6}

kilo (k) = 1,000 = 10^{3}

deci (d) = 0.1 = 10^{-1}

centi (c) = 0.01 = 10^{-2}

milli (m) = 0.001 = 10^{-3}

micro (μ) = 0.000,001 = 10^{-6}

nano (n) = 0.000,000,001 = 10^{-9}

pico (p) = 0.000,000,000,001 = 10^{-12}

femto (f) = 0.000,000,000,000,001 = 10^{-15}

- **Copyright and Disclaimer:**

No scientists were harmed during the making of this book.

Made in the
USA
Middletown, DE

77375986R00080